Code Division Multiple Access (CDMA)

Code Division Multiple Access (CDMA)

R. Michael Buehrer

www.morganclaypool.com

ISBN: 1598290401   paperback
ISBN: 9781598290400   paperback

ISBN: 159829041X   ebook
ISBN: 9781598290417   ebook

DOI 10.2200/S00017ED1V01Y200508COM002

A Publication in the Morgan & Claypool Publishers series
SYNTHESIS LECTURES ON COMMUNICATIONS #2

Lecture #2
Series Editor: William Tranter, Virginia Tech

Series ISSN: 1932-1244   print
Series ISSN: 1932-1708   electronic

First Edition
10 9 8 7 6 5 4 3 2 1

Printed in the United States of America

Code Division Multiple Access (CDMA)

R. Michael Buehrer

Virginia Polytechnic Institute and State University,
Blacksburg, Virginia, USA

SYNTHESIS LECTURES ON COMMUNICATIONS #2

 MORGAN & CLAYPOOL PUBLISHERS

ABSTRACT

This book covers the basic aspects of Code Division Multiple Access or CDMA. It begins with an introduction to the basic ideas behind fixed and random access systems in order to demonstrate the difference between CDMA and the more widely understood TDMA, FDMA or CSMA. Secondly, a review of basic spread spectrum techniques is presented which are used in CDMA systems including direct sequence, frequency-hopping, and time-hopping approaches. The basic concept of CDMA is presented, followed by the four basic principles of CDMA systems that impact their performance: interference averaging, universal frequency reuse, soft handoff, and statistical multiplexing. The focus of the discussion will then shift to applications. The most common application of CDMA currently is cellular systems. A detailed discussion on cellular voice systems based on CDMA, specifically IS-95, is presented. The capacity of such systems will be examined as well as performance enhancement techniques such as coding and spatial filtering. Also discussed are Third Generation CDMA cellular systems and how they differ from Second Generation systems. A second application of CDMA that is covered is spread spectrum packet radio networks. Finally, there is an examination of multi-user detection and interference cancellation and how such techniques impact CDMA networks. This book should be of interest and value to engineers, advanced students, and researchers in communications.

KEYWORDS

CDMA, Multiple Access, Spread Spectrum, Multiuser Detection, TDMA, FDMA, Packet Radio Networks

This book is dedicated to those who patiently waited for me to finish: My wife Andrea and our children, Faith, JoHannah, Noah, Gabrielle and Ruth.

Contents

1. **Multiuser Communications** 1
   1.1 Conflict-Free Medium Access Control 2
       1.1.1 Time Division Multiple Access 3
       1.1.2 Frequency Division Multiple Access 5
       1.1.3 Code Division Multiple Access 6
       1.1.4 Traffic Engineering and Trunking Efficiency 8
       1.1.5 Frequency Reuse 9
   1.2 Contention-Based Medium Access Control 15
       1.2.1 ALOHA 16
       1.2.2 Carrier Sense Multiple Access and Carrier Sense Multiple
             Access/Collision Avoidance 16
       1.2.3 Other Random Access Methods 19
   1.3 Multiple Access with Spread Spectrum 21
   1.4 Summary 22

2. **Spread Spectrum Techniques for Code Division Multiple Access** 23
   2.1 Forms of Code Division Multiple Access 23
   2.2 Direct Sequence Code Division Multiple Access 24
       2.2.1 Power Spectral Density of Direct Sequence Spread Spectrum 27
       2.2.2 Multiple Access 30
   2.3 Frequency Hopping 35
       2.3.1 Slow Versus Fast Hopping 38
       2.3.2 Power Spectral Density of Frequency-Hopped
             Spread Spectrum 39
       2.3.3 Multiple Access 41
   2.4 Time Hopping 43
   2.5 Link Performance of Direct Sequence Spread Spectrum in Code
       Division Multiple Access 45
       2.5.1 Additive White Gaussian Noise 45
       2.5.2 Multipath Fading Channels 46
       2.5.3 Impact of Bandwidth 54

2.6   Multiple Access Performance of Direct Sequence Code
        Division Multiple Access..56
        2.6.1   Gaussian Approximation 57
        2.6.2   Improved Gaussian Approximation 64
2.7   Link Performance of Frequency-Hopped Spread Spectrum...................64
2.8   Multiple Access Performance of Frequency-Hopped
        Code Division Multiple Access 67
2.9   Summary...71

3. **Cellular Code Division Multiple Access** 73
3.1   Principles of Cellular Code Division Multiple Access 73
        3.1.1   Interference Averaging 73
        3.1.2   Frequency Reuse..77
        3.1.3   Soft Hand-Off 79
        3.1.4   Statistical Multiplexing..82
3.2   Code Division Multiple Access System Overview 83
3.3   Capacity..85
        3.3.1   Comparison of Multiple Access Capacity 85
        3.3.2   Second-Order Analysis...87
        3.3.3   Capacity–Coverage Trade-Off 93
        3.3.4   Erlang Capacity 95
3.4   Radio Resource Management 100
        3.4.1   Power Control...100
        3.4.2   Mobile-Assisted Soft Hand-Off 105
        3.4.3   Admission Control 107
        3.4.4   Load Control 109
3.5   Summary...110

4. **Spread Spectrum Packet Radio Networks** 111
4.1   Code Assignment Strategies 112
        4.1.1   Common-Transmitter Protocol 112
        4.1.2   Receiver-Transmitter Protocol 116
4.2   Channel Access Strategies 117
4.3   Direct Sequence Packet Radio Networks 118
4.4   Frequency-Hopped Packet Radio Networks 120
        4.4.1   Perfect Side Information 121
        4.4.2   No Side Information.. 124
4.5   Summary...124

5.   Multiuser Detection...**127**
    5.1   System Model 127
    5.2   Optimal Multiuser Reception 129
    5.3   Linear Sub-Optimal Multiuser Reception 131
        5.3.1   The Decorrelating Detector 132
        5.3.2   Linear Minimum Mean Squared Error Receiver 137
    5.4   Non-Linear Sub-Optimal Receivers: Decision Feedback 137
        5.4.1   Decorrelating Decision Feedback 138
        5.4.2   Successive Interference Cancellation 140
        5.4.3   Parallel Interference Cancellation 147
        5.4.4   Multistage Receivers 149
    5.5   A Comparison of Sub-Optimal Multiuser Receivers 156
        5.5.1   AWGN Channels 156
        5.5.2   Near-Far Performance 157
        5.5.3   Rayleigh Fading 160
        5.5.4   Timing Estimation Errors 161
    5.6   Application Example: IS-95 162
        5.6.1   Parallel Interference Cancellation 165
        5.6.2   Performance in an Additive White Gaussian Noise Channel 167
        5.6.3   Multipath Fading and Rake Reception 168
        5.6.4   Voice activity, power control, and coding 168
        5.6.5   Out-of-Cell Interference 170
    5.7   Summary 171

Preface

The objective of this book is to provide the reader with a concise introduction to the use of spread spectrum waveforms in multiple user systems, often termed code division multiple access or CDMA. The book has been an outgrowth of course notes presented in a graduate-level course on spread spectrum communications. This book should provide sufficient material to cover CDMA systems within a graduate-level course on spread spectrum, advanced digital communication, or multiple access. The text should also be useful for working engineers who desire a basic understanding of the fundamental concepts underlying CDMA.

The reader of this book is expected to have a fundamental understanding of digital communications and some understanding of wireless systems in general. Additionally, readers are assumed to be generally familiar with basic stochastic processes, detection theory, and communication theory. The book builds on these fundamentals by explaining how spread spectrum systems differ from standard digital communication systems and, more importantly, how spread spectrum waveforms can be used as a means of channelization in a multiple user scenario.

The book covers the basic aspects of CDMA. In Chapter 1, the basic ideas behind conflict-free and contention-based systems are introduced to demonstrate the difference between CDMA and more widely understood orthogonal access techniques such as time division multiple access and frequency division multiple access. Additionally, random access schemes such as carrier sense multiple access are examined.

In Chapter 2, basic spread spectrum techniques that are used in CDMA systems are reviewed, including direct sequence spread spectrum, frequency-hopped spread spectrum, and time-hopping approaches. Both the link performance of such waveforms (in additive white Gaussian noise channels and fading channels) as well as the multiple access performance are examined. Special emphasis is given to fading channels since spread spectrum is more advantageous in these channels.

Once the basic concept of CDMA is presented, Chapter 3 focuses on cellular CDMA systems. Specifically, four basic principles of cellular CDMA systems are presented, and their impact on the performance of CDMA is explained. These four basic concepts include interference averaging, universal frequency reuse, soft hand-off, and statistical multiplexing. While the discussion is general, the CDMA cellular standard IS-95 is often used as an example. Additionally, important CDMA system functions (often termed radio resource management techniques), such as power control, mobile-assisted hand-off, load control, and admission control, are examined. Finally, the capacity of CDMA cellular systems on both the uplink and downlink is derived, emphasizing the differences between the links.

Spread spectrum waveforms are used not only in fixed access techniques such as in cellular systems, but also for multiple access in packet radio networks (PRNs). Chapter 4 discusses spread spectrum based PRNs emphasizing the differences between PRNs and their better known cellular counterparts. A primary emphasis is on spreading code assignment techniques, which is crucial in non-centralized systems.

Finally, Chapter 5 focuses on multiuser detection. A primary limitation of CDMA link performance and system capacity is in-cell multiple access interference (MAI). Multiuser detection is one means of mitigating MAI on the uplink of CDMA systems. The discussion of multiuser detection algorithms is broken down into two basic categories: linear techniques and non-linear techniques. Linear techniques discussed include the decorrelating detector and the minimum mean square error detector. Among the non-linear approaches examined, parallel interference cancellation and successive interference cancellation are the most prominent. Finally, all these techniques are compared, the benefits and detriments of each approach are mentioned, and the application of multiuser detection to the IS-95 cellular standard is examined.

I would like to acknowledge the many students who helped make this book possible. First, I would like to thank all of the students who have taken my spread spectrum course for their comments and input. I would also like to specifically thank several graduate students who helped with various plots and simulations including (but certainly not limited to) Ihsan Akbar, Dan Hibbard, Jihad Ibrahim, Nishant Kumar, Rekha Menon, and Swaroop Venkatesh as well as former colleagues at Bell Labs, especially Steve Nicoloso and Rob Soni for their assistance in Section 5.6. I also want to thank Lori Hughes for her many hours of editing that (or is it "which"?) greatly improved the initial manuscript.

CHAPTER 1

Multiuser Communications

All communication systems that support multiple users must have a set of protocols to allow these multiple users to share a common access medium. This single-access medium may be explicitly shared or broken up into smaller pieces, termed *channels*, that must also be shared. A logical channel can be defined as some fraction of the available access medium that is used by a particular transmit/receive pair. The system may contain only a single channel or tens of channels. However, in either case, the number of possible transmit/receive pairs in the system typically far outweighs the number of channels that are available. Thus, communication systems must have some mechanism for sharing the available channels among active transmit/receive pairs. This mechanism is termed *multiple access control* or sometimes *medium access*. Many methods exist for providing multiple access, and we will briefly examine the major techniques in this chapter. However, the focus of this book is one specific multiple access technique termed *code division multiple access* or CDMA, which is inherently associated with spread spectrum communication techniques. Since spread spectrum was developed as a military technology, CDMA techniques traditionally have been limited to military systems [1]. However, since the early 1990s, CDMA has been heavily investigated for commercial systems. Of particular note is the success that CDMA has experienced in commercial cellular systems. The first commercial CDMA standard, IS-95, pioneered by Qualcomm, Inc., was one of the three major second generation cellular standards and led to a third generation of cellular systems that was dominated by CDMA techniques (at least for voice services). However, before we discuss CDMA in detail, we must first discuss the more general concept of multiple access.

Two basic classes of multiple access are illustrated in Figure 1.1 [2]: contention-based techniques and conflict-free techniques. Conflict-free multiple access involves some type of reservation scheme in which the resources are divided into multiple channels. These channels are then reserved (through some other mechanism) for use by transmit/receive pairs for the duration of their communication. On the other hand, contention-based multiple access techniques allow no reservation. Instead, users must contend for the system resources whenever communication takes place. Such systems typically use the entire access medium as a single channel, although multi-channel versions are certainly possible. Both types of systems have several variations and

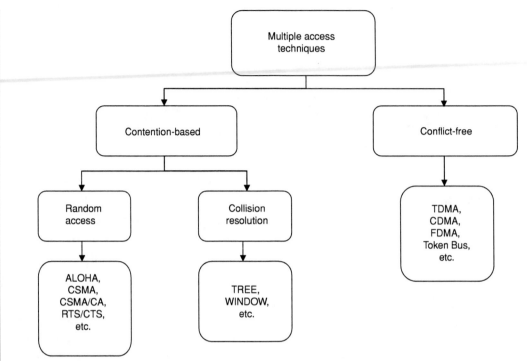

FIGURE 1.1: Multiple access techniques.

provide benefits for different types of traffic. We will now discuss the two major classes of multiple access.

1.1   CONFLICT-FREE MEDIUM ACCESS CONTROL

As mentioned previously, contention-free (or conflict-free) multiple access involves the division of system resources (i.e., the access medium) into fixed channels, which are then reserved by transmit/receive pairs for communication. In this way, users are guaranteed a channel for the duration of their communication. This type of multiple access is particularly beneficial for applications that require continuous, regular access to a channel, such as voice or video. However, for bursty data sources, such a scheme is inefficient because the channel is very often unused while it is reserved. The main difference between the types of contention-free multiple access is in how the channels are defined. In time division multiple access (TDMA), channels are defined according to time slots. In frequency division multiple access (FDMA), channels are defined according to frequency bands, and in CDMA, channels are defined not by time or frequency but by a spread spectrum parameter known as a *spreading code*. We will briefly review each type.

1.1.1 Time Division Multiple Access

TDMA systems define channels according to time slot. In other words, system time is defined as a series of repeating, fixed-time intervals (often called *frames*) that are further divided into a fixed number of smaller time periods called *slots*. When a transmit/receive pair is given permission to communicate, it is assigned a specific time slot in which to do so. Every time frame, each transmit/receive pair may communicate during its slot. An example is given in Figure 1.2 for four time slots. Typically, all users are given an opportunity to transmit once during a frame. Thus, the total frame is made up of K user slots and K guard times where K is the number of transmitters actively accessing the medium or equivalently the number of channels. Guard times are inserted to prevent collisions due to imperfect synchronization. The user throughput is a function of the overall system transmission rate and the number of time slots available (i.e., the fraction of time they are permitted to transmit).

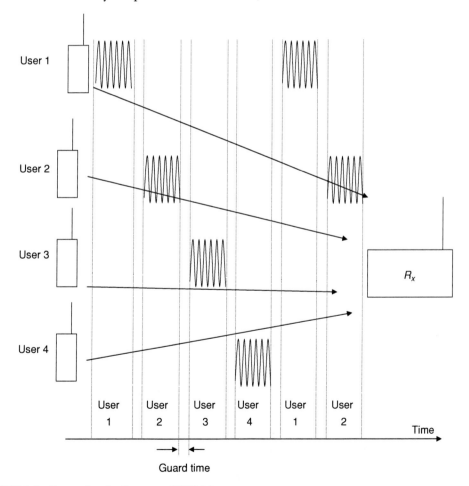

FIGURE 1.2: Example of a four-user TDMA system.

The example in Figure 1.2 demonstrates a centralized system in which multiple users communicate to a single receiver.[1] A decentralized system can also use TDMA, but providing strict time synchronization in a large, decentralized system can be very challenging. Additionally, the example given shows only one side of the communication (mobile to base station) and inherently assumes frequency division duplex (FDD) operation in which the channels from the centralized transmitter to the distributed receivers occur on a different frequency channel. This second band is also divided into time slots for transmission to the separate users. However, TDMA systems can also use time division duplex (TDD) in which time is broken into two consecutive frames; the first time frame is used for uplink (or downlink) transmission, and the second time frame is used for downlink (or uplink) transmission.

In a pure TDMA system, each transmitter occupies the entire bandwidth when transmitting. The system bit rate R_b^S is the rate at which each user transmits when accessing the channel. Ignoring guard times, the data rate per user is $R_b = (R_b^S)/K$, where K is the number of time slots per frame. If guard times are included, the relationship is a little more complicated. The time allocated per channel is simply equal to $(T_f/K) - T_g$ where T_f is the frame duration, K is the number of time slots (i.e., channels), and T_g is the guard time. The data rate per channel is equal to the number of bits transmitted per user divided by the frame duration. Thus,

$$R_b = \frac{\left(\frac{T_f}{K} - T_g\right) R_b^S}{T_f}$$

$$= \frac{R_b^S}{K} - \frac{T_g}{T_f} R_b^S \tag{1.1}$$

Clearly, we wish to have a small guard time to improve the efficiency of the system. However, practical considerations limit the minimum size of T_g. The bandwidth of the system is proportional to the system data rate:

$$B_S \propto \frac{R_b^S}{k} \alpha \tag{1.2}$$

where k is the number of bits per symbol and α is a constant related to the filtering, pulse shape, and modulation scheme.

There are advantages and disadvantages of TDMA as compared to other multiple access schemes. One advantage of the scheme is that it requires only a single radio frequency/intermediate frequency (RF/IF) section since all channels have the same frequency characteristics. Another advantage of TDMA is the ease with which variable data rates and asymmetric

---

[1]Note that the transmission from distributed users to a single receiver is typically referred to as the *uplink* while transmission from the centralized transmitter to the distributed receivers is termed the *downlink*.

links are accommodated. Variable data rates can be assigned by simply assigning multiple time slots to a single transmit/receive pair. Asymmetric links can be accommodated by changing the relative duration of uplink and downlink time slots. The relationship between the user-specific data rate and the overall system data rate is further illustrated in the following example.

Example 1.1. A TDMA system is to be designed with ten channels and a guard time of 50µs. If quadrature phase shift keying (QPSK) modulation is used, what system symbol rate is needed to achieve a data rate of 200kbps with a frame duration 10ms?

Solution: The frame duration T_f is 10ms. Due to guard time, the total time available for transmission is $10\text{ms} - 10 * 50\text{µs} = 9.5\text{ms}$. The transmission time per channel is thus

$$\frac{9.5\text{ms}}{10} = 0.95\text{ms} \tag{1.3}$$

The required system data rate is then

$$R_b^S = \frac{10\text{ms} * 200\text{kbps}}{0.95\text{ms}} = 2.105\text{Mbps} \tag{1.4}$$

Alternatively, from (1.1),

$$R_b^S = \frac{R_b}{\frac{1}{K} - \frac{T_g}{T_f}}$$

$$= \frac{200\text{ kbps}}{0.1 - 0.005}$$

$$= 2.105\text{ Mbps} \tag{1.5}$$

Using QPSK (two bits per symbol), the symbol rate needed is $2.105/2 = 1.05\text{Msps}$.

1.1.2 Frequency Division Multiple Access

The second major type of contention-free multiple access is FDMA in which channels are defined according to frequency allocation. Thus, all transmitters are active simultaneously but occupy different segments of the RF spectrum as illustrated in Figure 1.3. In an FDMA system, the bandwidth per user is simply related to the data rate and modulation scheme used. The total bandwidth of the system is $B_S = K * B$ where K is the number of channels and $B = \alpha(R_b/k)$ is the bandwidth per channel ignoring guard bands. With guard bands, we have

$$B_S = K * (B + B_g) \tag{1.6}$$

where B_g is the size of the guard band.

The efficiency of TDMA and FDMA are essentially the same, with slight differences depending on the guard times/bands required. Both techniques are referred to as *orthogonal*

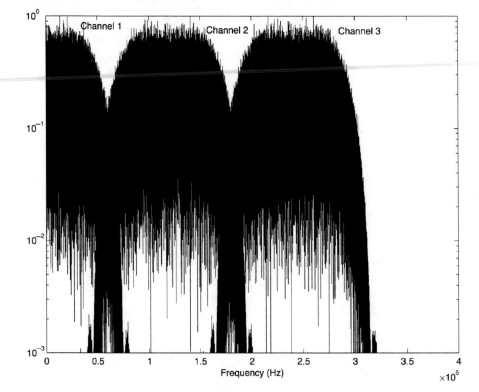

FIGURE 1.3: Example of three-channel FDMA system.

multiple access techniques since, ideally, there is no interference between channels. An advantage of FDMA over TDMA is the substantial reduction in the required symbol rate. Another advantage of the FDMA is that no tight synchronism between users is required and strict isolation between channels is relatively easy to maintain with properly designed filters and tuners. However, disadvantages include the fact that with a larger number of channels the IF filter (used to select the channel of interest) must be fairly narrow and allocating variable data rates requires multiple receive filters.

1.1.3   Code Division Multiple Access

As we have seen from the previous two sections, a key to contention-free multiple access is the definition of orthogonal channels. Orthogonal channels are channels in which a system user can communicate without causing interference to another user. The orthogonality can be created either in the time domain or in the frequency domain. In TDMA, users simply transmit at different times, thus maintaining orthogonality in the time domain. In FDMA, users transmit in different frequency bands, which creates orthogonality in the frequency domain since receivers can filter out unwanted frequency bands. In reality, the channels in these access techniques are not truly orthogonal due to imperfect filters or imperfect time synchronization, but the cross-channel interference is very small.

In CDMA systems, channels are defined not by time or frequency but by code. Spread spectrum systems (as we will see in more detail in Chapter 2) rely on pseudo-random waveforms termed *spreading codes* to create noise-like transmissions. If users can be given different codes that have low cross-correlation properties, channels can be defined by those codes. To better understand this, let us consider an FDMA system with two channels. Assuming a linear modulation scheme, the transmit signals from two distinct users can be written as

$$s_1(t) = \sqrt{2P_1}b_1(t)\cos(\omega_1 t + \theta_1) \qquad (1.7a)$$

$$s_2(t) = \sqrt{2P_2}b_2(t)\cos(\omega_2 t + \theta_2) \qquad (1.7b)$$

where P_i, $b_i(t)$, ω_i, and θ_i are the transmit power, data signal, transmit frequency, and random phase offset for the ith user, respectively. Now consider the received signal (normalized so that the desired signal is received at maximum power) at user 1:

$$r_1(t) = s_1(t) + \sqrt{PL_R}s_2(t) + n(t) \qquad (1.8)$$

where PL_R is the relative path loss between transmitter 2 and receiver 1 and $n(t)$ is additive white Gaussian noise (AWGN). Now, assuming square pulses for simplicity, the output of the matched filter receiver at user 1 for an arbitrary symbol period is

$$Z = \frac{1}{T}\int_0^T r_1(t)\cos(\omega_1 t + \theta_1)\, dt \qquad (1.9)$$

where T is the data symbol period and we have examined the first symbol period. Provided that $|\omega_1 - \omega_2| \gg T$, the signal $s_2(t)$ will produce no response to the filter matched to signal $s_1(t)$. Thus, we can say that the channels are orthogonal. We can write similar equations for TDMA where the channels are defined by a time slot. However, in CDMA, the channels are defined by spreading codes. For example, with direct sequence CDMA (DS-CDMA), the two signals can be defined by

$$s_1(t) = \sqrt{2P_1}a_1(t)b_1(t)\cos(\omega_1 t + \theta_1) \qquad (1.10a)$$

$$s_2(t) = \sqrt{2P_2}a_2(t)b_2(t)\cos(\omega_1 t + \theta_2) \qquad (1.10b)$$

where $a_1(t)$ and $a_2(t)$ are spreading codes that define the "channel" for each user signal. Thus, the cross-correlation between $a_1(t)$ and $a_2(t)$ dictates the performance of CDMA as we shall see. In other words, we wish

$$\rho_{12} = \frac{1}{T}\int_0^T a_1(t)a_2(t)\, dt \qquad (1.11)$$

to be small to avoid excessive interference between users. In general, will be demonstrated $|\rho_{ij}| > 0$, $\forall i \neq j$, and thus, CDMA is a non-orthogonal multiple access scheme. However, CDMA has other advantages as we shall see.

1.1.4   Traffic Engineering and Trunking Efficiency

The previous discussion focused on the number of channels supported by a specific system. However, in most systems, the number of users that needs to be supported typically far exceeds the number of channels available. However, since not all users of the system require the use of a channel at the same time, a small number of channels can statistically support a large number of users. This concept is known as *trunking* and exploits the statistical behavior of user access. Determining the user population that can be supported by a number of channels is termed *traffic engineering* or *traffic analysis*. Classic traffic analysis is based on voice applications and was originally developed to determine the number of telephone circuits required to handle a given number of telephones [3]. Clearly, due to the random nature of users accessing the system, some probability exists such that the number of users requesting a channel will exceed the number of channels available. When this happens, we say that an attempt to use the system is *blocked*. For cellular voice systems, the probability that a user is blocked during the busiest hour of the day is termed the *grade of service* (GOS).

The probability of blocking is determined using a queuing model for the system [4]. To develop this queuing model, it is typically assumed that the time between user requests for a channel is exponentially distributed (i.e., short times between consecutive requests are more likely than very long times between consecutive requests). Such an assumption leads to a Poisson distribution for the number of requests in a given time frame based on some average arrival rate λ. It is further assumed that the service time (i.e., the length of time that the channel is required by the user) is also exponentially distributed with an average service time $1/\mu$. Since there is a finite number of channels K available, this scenario is modeled by the classic *M/M/K/K* queue [4]. Since we further assume that blocked calls are cleared (as opposed to simply being delayed for some amount of time), the probability of blocking is simply the probability that all K channels are being used and is termed the *Erlang B formula*, which is given by

$$\Pr\{\text{blocking}\} = \frac{\frac{\Lambda^K}{K!}}{\sum_{k=0}^{K} \frac{\Lambda^k}{k!}} \qquad (1.12)$$

where $\Lambda = \lambda/\mu$ is the offered traffic load in Erlangs. As an example, with $K = 20$ channels and an offered load of $\Lambda = 13.2$ Erlangs, the blocking probability is calculated as 2%. Assuming that the average user traffic is 0.02 Erlangs/subscriber (i.e., each subscriber makes one 72-second call per hour), the total number of subscribers that can be supported with a 2% blocking probability is $13.2/0.02 = 660$.

Example 1.2. A specific service area has 850 subscribers. Assuming that the average user traffic is 0.05 Erlangs/subscriber, determine the minimum number of channels necessary to guarantee no worse than a 1% blocking probability. If the service area were doubled such that 1700 subscribers were supported (assuming a uniform user density), how many channels are needed?

Solution: Given that there are 850 subscribers and 0.05 Erlangs/subscriber, the system must support

$$\Lambda = 850 * 0.05$$
$$= 42.5 \text{ Erlangs} \qquad (1.13)$$

We must thus solve (1.12) for K with $\Lambda = 42.5$ and Pr{blocking} = 0.01. This is simplified by using an Erlang B table as provided in Appendix A. From Table A.1, we find that we require $K = 56$ channels. If the number of subscribers were doubled ($\Lambda = 85$ Erlangs), we would require $K = 102$ channels.

Note that in the preceding example, when doubling the number of subscribers supported, the number of channels required does not double but increases by a factor of less than two. This effect is known as *trunking efficiency* and is a general principle of queueing systems. By increasing the pool of channels available by a factor of X, the number of subscribers supported is increased by more than X due to the statistical nature of channel usage.

1.1.5   Frequency Reuse

Even with trunking efficiency, the number of channels required to cover a large metropolitan area (e.g., in a cellular system) is quite large. For example, in a city of two million people with a 10% market penetration rate, if a service provider has 20% market share, it must support 40,000 subscribers. Even if each subscriber generates only 0.02 Erlangs, the total system load is 800 Erlangs, leading to a large number of required channels. To deal with the large number of required channels, frequency reuse is used in cellular systems as it is in many wireless systems, including broadcast radio and television, wireless local area networks (WLANs), and public safety radio networks. However, the frequency reuse concept gained most of its popularity through its use in cellular systems, and it is illustrated in Figure 1.4. Because transmitted signal power decays exponentially with distance, co-frequency channels (whether TDMA or FDMA) can be reused in geographically separated locations. Typically, a set of K channels is divided into Q sets where Q is the reuse factor and contiguous areas utilize one of the Q sets. A given set of channels can be used in two areas if the areas are separated by sufficient distance to ensure that two channels using the same time slot and/or frequency avoid interfering with each other. Figure 1.4 demonstrates an example of a reuse factor of $Q = 7$. Note that a tessellation

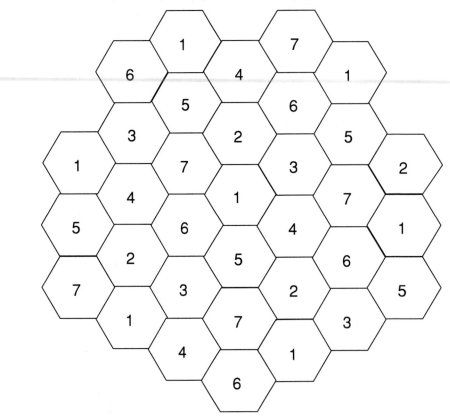

FIGURE 1.4: Classic frequency reuse for a reuse pattern of 7.

of hexagons is used to cover the entire service area. This theoretical coverage pattern in which individual geographic areas are called *cells* is the origination of the term *cellular* [5].

To ensure a minimum distance between co-frequency cells, only certain frequency reuse patterns are possible. Specifically, $Q = i^2 + ij + j^2$ for any positive integers i and j [6]. With such patterns, it can be shown that the minimum distance between co-frequency cells is $D = \sqrt{3Q}d_r$ where d_r is the cell radius [6]. This distance along with the maximum interference tolerable determines the allowable reuse factor.

The reuse of channels means that co-channel interference is received by each receiver in the system. The reuse pattern used depends on the minimum signal-to-interference ratio (SIR) that can be tolerated. The SIR experienced in a system depends on the geography of the area, the building size and density, and other environmental factors as well as the reuse pattern. However, for the sake of discussion, let us assume that the environment is uniform, all cells have the same size, and the transmit power decays with d^κ where d is the distance from the transmitter to the receiver and κ is termed the *exponential path loss factor*. It can be shown that six cells are always in the first tier of co-channel cells as seen in Figure 1.4. If the interference from the first tier dominates the interference, the SIR experienced can be calculated as

$$SIR = \frac{P_r}{\sum_{k=1}^{6} I_k} \tag{1.14}$$

where P_r is the desired received signal power and I_k is the received signal power from the kth co-channel interferer. The worst SIR on the downlink will occur when the mobile is at the cell edge, i.e., $d = d_r$. It can be shown that the uplink (the link from the mobile to the base station) and downlink (the link from the base station to the mobile) provide similar results, so we will look only at the downlink. Thus, if P_t is the transmit power at each base station, the SIR is

$$\begin{aligned}
SIR &= \frac{\frac{P_t}{d_r^{\kappa}}}{\sum_{k=1}^{6} \frac{P_t}{D_k^{\kappa}}} \\
&= \frac{d_r^{-\kappa}}{6 D^{-\kappa}} \\
&= \frac{d_r^{-\kappa}}{6 \left(\sqrt{3Q} d_r\right)^{-\kappa}} \\
&= \frac{\left(\sqrt{3Q}\right)^{\kappa}}{6}
\end{aligned} \tag{1.15}$$

where D_k is the distance of the kth interfering base station to the mobile of interest and we use the approximation that $D_k \approx D, \forall k$. Thus, we can see that an increase in Q provides an improvement in SIR. However, for a fixed number of cells, increasing Q decreases the number of channels available per cell. Thus, a trade-off exists between the required SIR and spectral efficiency as shown in the following examples.

Example 1.3. Assuming a path loss factor of $\kappa = 4$, determine the maximum number of channels per cell if there are 450 total channels available and the required SIR is 18dB. Does the answer change if you include both the first and the second tier of co-channel interferers?

Solution: From (1.15), we have

$$10^{1.8} \leq \frac{\left(\sqrt{3Q}\right)^4}{6} \tag{1.16}$$

Rearranging, we have

$$\begin{aligned}
Q &\geq \frac{\sqrt{6 * 10^{1.8}}}{3} \\
&= 6.48
\end{aligned} \tag{1.17}$$

The smallest valid value of Q greater than 6.48 is then $Q = 7$ ($i = 2$, $j = 1$). Thus, the maximum number of allowable channels per cell is

$$K = \frac{450}{7}$$
$$= 64 \tag{1.18}$$

Now, if we include the second tier of interferers, we have

$$SIR = \frac{\frac{P_t}{d_r^\kappa}}{\sum_{k=1}^{6} \frac{P_t}{D_k^\kappa} + \sum_{k=7}^{12} \frac{P_t}{D_k^\kappa}} \tag{1.19}$$

It can be shown that $D_7 = D_8 = \cdots = D_{12} = 2D$. Thus, we have

$$SIR = \frac{d_r^{-\kappa}}{6\left(\sqrt{3Q}d_r\right)^{-\kappa} + 6\left(2\sqrt{3Q}d_r\right)^{-\kappa}}$$
$$= \frac{\left(\sqrt{3Q}\right)^\kappa}{6 + 6/2^\kappa} \tag{1.20}$$

which leads to

$$Q = \frac{\sqrt{\left(6 + 6/2^4\right) * 10^{1.8}}}{3}$$
$$= 6.68 \tag{1.21}$$

Clearly, including the second tier of interferers makes no significant difference, and we are justified in ignoring it.

Frequency reuse is heavily dependent on propagation conditions as well as the desired SIR as we can see in the following example.

Example 1.4. Repeat Example 1.3 if the path loss factor is $\kappa = 3$.

Solution: Repeating the analysis from Example 1.3 with $\kappa = 3$, we have

$$Q = \frac{\left(6 * 10^{1.8}\right)^{2/3}}{3}$$
$$= 17.4 \tag{1.22}$$

The smallest valid value of Q greater than 17.4 is $Q = 19$ ($i = 3$, $j = 2$), and the maximum number of channels per cell is $K = 23$. Thus, we see that while a larger path loss factor means that more power is required to cover a particular area (i.e., there is more path loss at a fixed distance), a larger path loss factor actually benefits capacity in a multi-cell scenario since greater isolation between cells is experienced.

The previous example showed that the efficiency of frequency reuse improves as the propagation conditions worsen. In the next example, we show that the efficiency is also heavily dependent on the desired SIR. Thus, if the system can tolerate lower values of SIR, the overall system efficiency can be improved.

Example 1.5.   Repeat Example 1.3 if an SIR of 12dB can be tolerated.

Solution: If an SIR of 12dB can be supported, we have (assuming $\kappa = 4$),

$$Q = \frac{\sqrt{6 * 10^{1.2}}}{3}$$
$$= 3.25 \qquad\qquad (1.23)$$

The smallest allowable value of Q greater than 3.25 is $Q = 4$, which provides $K = 450/4 = 112$ channels per cell. Thus, we clearly see that if we can tolerate lower SIR, we can increase our capacity (i.e., the number of channels per cell).

It should be noted that frequency reuse is classically associated with FDMA. Theoretically, there is no reason why pure TDMA cannot also employ reuse, although, practically speaking, synchronization across multiple cells would pose a significant practical challenge. If systems employ some combination of FDMA and TDMA, frequency bands can be divided according to cells and reused as is done in second generation cellular systems based on the standards IS-136 and Global System for Mobile Communications (GSM) [7]. CDMA, however, does not typically employ reuse patterns. In fact, the use of universal frequency reuse (i.e., a reuse pattern of 1) is a significant advantage of CDMA as we will discuss in detail in Chapter 3. To demonstrate the difference between the interference statistics of FDMA and CDMA systems, consider a TDMA/FDMA system with a reuse factor of $Q = 7$. As mentioned previously, the average path loss with distance in a wireless system can be written as

$$\overline{PL} \propto d^{\kappa} \qquad\qquad (1.24)$$

However, because of terrain and various buildings in the environment, path loss versus distance is typically found to be a log-normal random variable where the mean path loss is given as in (1.24) and the standard deviation is between 6 and 10dB [7]. This variation is termed *shadowing*. Thus, the SIR experienced on a particular link is a random variable depending on the location of the various mobiles and the shadowing experienced by each. Specifically, the

SIR including log-normal shadowing can be written as

$$SIR = \frac{P_r}{\sum\limits_{k=1}^{6} I_k} \qquad (1.25)$$

where P_r and I_k are log-normal random variables. If we assume that the system uses power control such that every uplink signal is received at its base station with the same power P, the SIR for an FDMA system can be rewritten as

$$SIR = \frac{P}{\sum\limits_{k=1}^{6} P\left(\frac{d_k}{D_k}\right)^{\kappa} 10^{(l_k - l'_k)/10}}$$
$$= \frac{1}{\sum\limits_{k=1}^{6} \left(\frac{d_k}{D_k}\right)^{\kappa} 10^{(l_k - l'_k)/10}} \qquad (1.26)$$

where d_k and D_k are the distances from the kth co-channel interferer to the base station that it is communicating with and the base station of interest, respectively, and l_k and l'_k are the log-normal shadowing factors from the kth co-channel interferer to its base station and the base station of interest, respectively. On the other hand, with power control, the SIR for a CDMA system can be written as

$$SIR = \frac{N}{(K-1) + \sum\limits_{k=1}^{6K} f\left(\left(\frac{d_k}{D_k}\right)^{\kappa} 10^{(l_k - l'_k)/10}\right)} \qquad (1.27)$$

where N is the spreading gain (i.e., the ratio of the bandwidth to the data rate), K is the number of users per cell, and $f(x)$ is a function that guarantees that users are associated with the base station having the smallest path loss. That is,

$$f(x) = \begin{cases} x & 0 \leq x \leq 1 \\ 0 & \text{otherwise} \end{cases} \qquad (1.28)$$

The details of the CDMA SIR equation will be derived in Chapter 3. For now, we simply use (1.27) to compare the SIR statistics for the two cases. Note that for CDMA systems, $Q = 1$ and thus $D_k = \sqrt{3}d_r$. A set of 10,000 random scenarios was simulated for uniformly distributed users, and the empirical cumulative distribution functions (CDFs) are plotted in Figure 1.5 where the number of users K in the CDMA system was adjusted to achieve the same average SIR. We can see that for the same average SIR, the distributions are very different. Specifically,

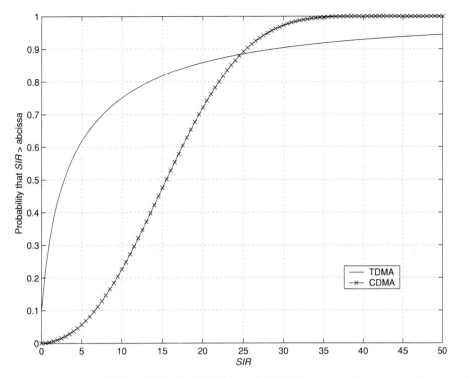

FIGURE 1.5: Empirical CDFs of SIR for FDMA and CDMA systems (both normalized to give an average SIR of 12dB).

the CDMA system exhibits very little spread in the SIR value compared to the FDMA system. This can be seen from the steep slope of the CDMA CDF plot. Since communication system performance depends on the tails of the SIR (or signal-to-noise ratio, SNR) distribution, the heavy tails of the SIR distribution in the FDMA case mean that the average SIR must be significantly higher to achieve the same 90% value. We will examine this more thoroughly in Chapter 3.

1.2   CONTENTION-BASED MEDIUM ACCESS CONTROL

Contention-free multiple access techniques are efficient provided that traffic is relatively continuous. If traffic is bursty, contention-free systems waste channels by dedicating them to a single transmit/receive pair. Instead, systems with bursty traffic typically use contention-based multiple access schemes. In contention-based schemes, the entire resource is dedicated to a single channel and all users must contend to use the channel when they need to transmit.

1.2.1   ALOHA

The most common contention-based methods are random access methods. The first random access method was developed by Abramson and is known as ALOHA [8,9]. In this technique, users attempt to access the channel whenever they have data to transmit. If two users transmit at the same time (or within a packet time), a collision occurs. When the receiver fails to acknowledge receipt of the transmission, the transmitter realizes that a collision has occurred and retransmits the packet. However, if the two transmitters whose packets collided both retransmitted as soon as they realized that a collision occurred, another collision would occur. Thus, the key to the random access scheme is that each transmitter waits a random period before retransmitting. This random back-off period decreases the probability of a second collision as seen in Figure 1.6.

While this technique is a useful means of allocating the channel when traffic is random and infrequent, it is inefficient. Specifically, the throughput of the ALOHA protocol can be shown to be

$$S = \lambda e^{-2\lambda} \qquad (1.29)$$

where λ is the arrival rate of packets per packet time. The throughput is plotted in Figure 1.7. An improvement in throughput can be realized if transmissions are synchronized so that the probability of collision is reduced by a factor of 2. This is termed *Slotted ALOHA* and the resulting throughput is also shown in Figure 1.7. We can see that by adding the additional structure to the random access, we can double the peak throughput. However, this requires network-wide synchronization, which can be difficult to achieve in practice.

1.2.2   Carrier Sense Multiple Access and Carrier Sense Multiple Access/Collision Avoidance

The main drawback to ALOHA and Slotted ALOHA is that transmitters blindly transmit without attempting to determine if the channel is in use. Carrier sense multiple access (CSMA) and carrier sense multiple access/collision avoidance (CSMA/CA) are both contention-based

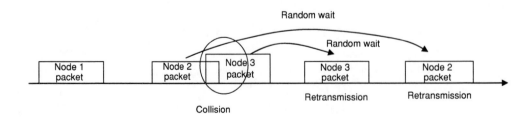

FIGURE 1.6: Illustration of the ALOHA random access protocol.

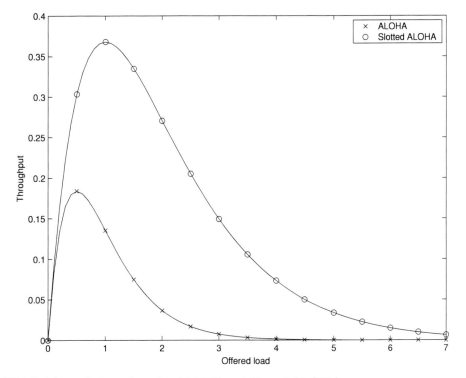

FIGURE 1.7: Network throughput for ALOHA and Slotted ALOHA.

medium access control (MAC) protocols that attempt to overcome this drawback. A node with a packet to transmit first senses the channel to check for an ongoing transmission—hence the term *carrier sense* (CS). If the node senses that the medium is free, it transmits its packet immediately. If it senses the medium is busy, it either waits until it is free and transmits (*persistent-1 CSMA*) or waits until it is free and then sets a random timer, waits for the timer to expire, and (if it has sensed no additional transmissions) then transmits (*non-persistent CSMA*). CSMA can also be slotted or unslotted just as ALOHA. The throughput of CSMA is plotted in Figure 1.8. Note that persistent-1 CSMA can provide better throughput than ALOHA and non-persistent CSMA at low loading levels. However, at high system loading factors, non-persistent CSMA provides far superior performance.

The previously described contention-based wireless networks suffer from the hidden node/exposed node problem. The hidden node problem is more severe than is the exposed node problem in most scenarios. The hidden node problem is demonstrated in Figure 1.9. The hidden node (Node 3) cannot sense the ongoing communication between the sender (Node 1) and the receiver (Node 2), senses the channel as idle, and proceeds with transmission of its packet to the

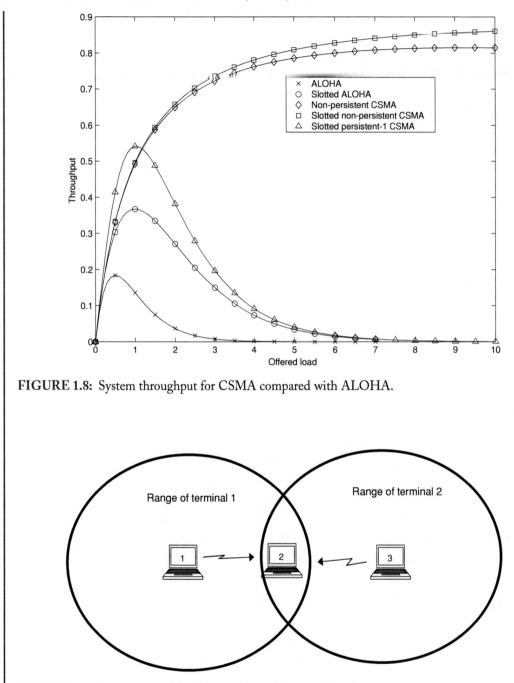

FIGURE 1.8: System throughput for CSMA compared with ALOHA.

FIGURE 1.9: Illustration of the hidden node problem in CSMA.

receiver, causing a collision at the receiver. The exposed node problem occurs when the exposed node senses the channel as busy because it can listen to the sender's ongoing communication with the receiver. The exposed node can still communicate with its intended receiver even if it senses a carrier because the proximity of a transmission to the transmitter does not necessarily indicate the proximity of a transmission to the receiver. Thus, it is possible that even though the transmitter suppressed transmission, it could have successfully communicated with its intended receiver. Both hidden and exposed node problems lead to a reduction in aggregate throughput. The CSMA protocol has no means to avoid the hidden node/exposed node problems.

To overcome the problem of the hidden and exposed terminals, the MACA (Multiple Access Collision Avoidance) protocol was proposed [10]. This protocol gets rid of the carrier sense in the CSMA protocol and instead uses a different algorithm for collision avoidance, hence the name MACA. Specifically, it relies on an RTS/CTS (request-to-send/clear-to-send) handshake to avoid collisions at the receiver. When Node A wishes to transmit to Node B, it first sends an RTS to Node B containing the length of the proposed data transmission. If the node hears the RTS and is not deferring, it replies with a CTS packet. When Node A hears the CTS, it immediately sends its data. Any node that hears the RTS defers all transmissions until after the expected reception of the CTS message. All nodes that hear the CTS message defer until the end of the data transmission. Thus, all nodes (and only those nodes) that are capable of interfering with the CTS or the data transmission avoid transmitting during the appropriate intervals.

The CSMA/CA protocol combines the carrier sense mechanism with collision avoidance. It solves the hidden node problem by using the RTS/CTS mechanism. This is sometimes termed a *virtual carrier sense*. Before making an attempt to send any data after the back-off interval associated with the collision avoidance has elapsed, the node again senses the channel. This technique helps resolve contention and reduces collision probability under high load conditions.

1.2.3 Other Random Access Methods
Most of the protocols discussed in the previous section require a particular node to listen for the carrier. It should be noted that carrier sense prevents collisions from happening at the transmitter, but most collisions occur at the receiver (the hidden node/exposed node problem as described previously). The lack of a carrier does not always indicate that it is safe to transmit (i.e., the hidden node problem), and the presence of a carrier does not always mean that the node should not transmit (i.e., exposed node problem). So carrier sense is not always an appropriate indication of the current channel utilization.

Bhargavan slightly modified the MACA protocol and proposed a new multiple access protocol termed MACAW (Multiple Access Collision Avoidance Protocol for Wireless LANs) [11]. This protocol proposed the addition of an ACK for every DATA packet sent. (This is now used in the 802.11 standards.) The ACK allows for quick determination of a lost packet. If an ACK is not received within a defined time frame, the transmitter assumes that the packet was lost and schedules a retransmission of the packet. This dramatically improves system throughput in noisy channels [11].

The protocol also adds a data-sending (DS) packet after the CTS message. The exchange sequence between the transmitter and the receiver thus looks like RTS–CTS–DS–DATA–ACK where the DS stands for the data-sending frame, which tells the nodes that a successful exchange of RTS/CTS has occured. This prevents an exposed node from attempting to transmit an RTS to a sender near to it, which would lead to large back-offs because the sender is already transmitting data to another node and would not respond to the exposed node's request.

Another variation of the MACA protocol is the MACA/BI (MACA—By Invitation) protocol first proposed by Talucci [12]. In this protocol, an RTS frame is not sent from an intended transmitter to the receiver. Instead, this is a receiver-initiated protocol in which the receiver determines when a sender is likely to send a packet (either by relying on the packet arrival rate or by the sender telling the receiver in the previous packet about a backlog of packets). The receiver then initiates (prepares the floor for transmission) a call by sending a CTS to the sender. The sender, after receiving the CTS, starts transmitting data to the receiver.

Tobagi also addressed the hidden node problem [13] by using a busy tone to indicate the ongoing transmission and thus preventing any other node from initiating another transmission. All the nodes monitor the busy tone to determine the availability of the channel. The proposed protocol does not use RTS and CTS for collision avoidance and depends on centralized access to avoid collisions. (By using a centralized access topology, channel access time is allocated to each user such that two nodes do not contend for the same channel time.) Attempts along similar lines were made [14,15] to avoid the hidden node problem. They also use the busy tone technique to avoid collisions.

The FAMA (Floor Acquisition Multiple Access) scheme was proposed [16] in which each node is required to acquire the channel before it may initiate the transmission. The node uses both carrier sensing and RTS/CTS to acquire the floor. Once the floor is acquired, the node can successfully transmit data. Fullmer studied FAMA/NPS (FAMA non-persistent packet sensing) and showed that packet sensing schemes alone could not solve the hidden node/exposed node problem [16]. FAMA was extended to FAMA/NCS (FAMA non-persistent carrier sensing), which uses a CTS dominance mechanism (longer CTS packets). If the node has begun transmission of the CTS packet and, at the same time, an RTS packet is sent, the node transmitting the RTS packet hears the CTS packet and refrains from accessing the channel.

Haas proposed dual busy tone multiple access (DBTMA) [17]. The protocol uses two out-of-band tones along with the RTS/CTS handshake for informing neighbors about an on-going transmission. The protocol resolves the hidden node/exposed node problem completely. A brief description of the algorithm is as follows. Once an RTS packet is transmitted, the BTt (busy tone—transmitter) signal is set to prevent the RTS from getting corrupted. On hearing the BTt tone, the other transmitters would refrain from sending an RTS packet and back-off. At the end of the RTS transmission, the transmitter turns off the BTt tone and waits for the CTS packet from the receiver. Once the RTS packet is received, the receiver responds with the CTS packet and sets the BTr (busy tone—receiver) signal. Any transmitter in the vicinity of the receiver hears the tone and does not transmit while the tone is set. It might happen that two simultaneous RTS packets are sent, which corrupts the RTS signal. In this case, the receiver would not understand the command and would not respond. Both the transmitters would individually time out and repeat the above procedure before again sending the RTS packet. This prevents corruption of the data. This algorithm also solves the hidden terminal/exposed terminal problem, as the hidden nodes can reply to RTS requests by setting their busy tones and the exposed node can initiate a transmission because it no longer needs to listen to the shared medium. Although the DBTMA scheme solves the hidden/exposed node problem, it requires two additional channels for setting the BTr and the BTt signals, which is a significant overhead in the already crowded spectrum allocated for WLANs.

The 802.11 MAC layer [18] is based on the CSMA/CA + ACK protocol for unicast frames and the CSMA/CD (carrier sense multiple access/collision detection) protocol for broadcast frames. It also deploys a virtual carrier sense mechanism (using RTS/CTS) to avoid a station from transmitting when two nodes are already communicating.

1.3 MULTIPLE ACCESS WITH SPREAD SPECTRUM

Theoretically, systems that utilize spread spectrum waveforms could use any of the multiple access schemes described earlier. Specifically, if the application lent itself to contention-free multiple access, a system could combine a spread spectrum physical layer with TDMA, FDMA, or CDMA. However, spread spectrum signals have a bandwidth N times larger than the data rate. The use of TDMA or FDMA would require N times as much bandwidth for the entire system. However, since in CDMA all signals can occupy the same spectrum, no additional bandwidth is needed to add more users. Thus, while spread spectrum signals are inefficient in terms of bandwidth, a CDMA system may have good bandwidth efficiency.

For contention-based multiple access, the previously-mentioned schemes (or adaptations thereof) can be used with spread spectrum waveforms. For example, some forms of 802.11 use direct sequence spread spectrum waveforms but use CSMA/CA for multiple access. We will discuss specific modifications of contention-based MAC protocols for spread spectrum in Chapter 4.

1.4   SUMMARY

In this chapter, we have investigated basic concepts in multiuser communications. Specifically, we have discussed fundamental techniques for allowing multiple pairs of users to communicate using the same medium. These techniques are typically divided into contention-free and contention-based techniques. Spread spectrum systems can use either type depending on the type of traffic in the system. Of primary importance in this book are CDMA techniques specifically for contention-free systems. In the following chapters, we will describe CDMA techniques more thoroughly for contention-free access as well as contention-based access schemes utilizing spread spectrum waveforms.

CHAPTER 2

Spread Spectrum Techniques for Code Division Multiple Access

In the previous chapter, we reviewed the basic concepts of multiuser communications and the multiple access techniques used to allow multiple users to communicate. In this chapter, we will focus on the major forms of spread spectrum communication and their application to CDMA. CDMA is based on spread spectrum techniques that originated in military communications. In CDMA, channels are defined by spreading waveforms or the spreading codes that underlie those waveforms. There are several types of spread spectrum, and thus there are several types of CDMA. We will focus on the two basic forms of CDMA: *direct sequence CDMA* (DS-CDMA) and *frequency-hopped CDMA* (FH-CDMA). We will also briefly mention a third type of CDMA termed *time-hopped CDMA* that is currently receiving attention for its application to ultra-wideband (UWB) systems.

2.1   FORMS OF CODE DIVISION MULTIPLE ACCESS

CDMA is also known as spread spectrum multiple access or SSMA because the use of spread spectrum waveforms is fundamental to CDMA.[1] Spread spectrum can be defined as any modulation technique that uses a bandwidth that is well beyond what is necessary for the data rate being transmitted and uses a pseudo-random signal to obtain the increased bandwidth. The latter factor distinguishes spread spectrum techniques from standard communication techniques such as frequency modulation (FM) and high-order orthogonal signaling, which may also require high bandwidth compared to the information rate. There are two main reasons why spread spectrum waveforms were traditionally used: low probability of intercept and resistance to jamming [1, 19]. These two properties are a direct result of both the excess bandwidth used by spread spectrum waveforms and the resulting low power spectral density (PSD) and can also be directly exploited to provide multiple access. In multiple access systems, we are concerned

---

[1]Some make a distinction between CDMA and SSMA in that CDMA specifically designs its spreading waveforms to have low cross-correlation properties, whereas SSMA systems have independent codes that may also have low cross-correlation [1]. We do not make such a distinction here.

with the interference from and to other spread spectrum waveforms rather than with hostile narrowband receivers or jamming signals.

There are two basic spread spectrum techniques: direct sequence spread spectrum (DS/SS) and frequency-hopped spread spectrum (FH/SS) [1, 19, 20]. These two techniques can be used for multiple access and are commonly termed DS-CDMA and FH-CDMA. We will examine both these techniques in the following sections (Sections 2.2 and 2.3) and discuss their performance in AWGN and fading channels as well as their multiple access capabilities. Both techniques rely on spreading waveforms to accomplish pseudo-random spreading. A key to CDMA is defining multiple spreading waveforms with low cross-correlation properties to allow multiple users to share the spectrum efficiently. A third technique that has gained more attention in recent years is termed time-hopped spread spectrum. This will be discussed briefly in Section 2.4. The link performance and multiple access capabilities of DS-CDMA will be discussed in Sections 2.5 and 2.6, respectively, and the link performance and multiple access capabilities of FH-CDMA will be discussed in Sections 2.7 and 2.8.

2.2   DIRECT SEQUENCE CODE DIVISION MULTIPLE ACCESS

DS/SS is perhaps the most common form of spread spectrum in use today. DS/SS accomplishes bandwidth spreading through the use of a high rate symbol sequence (termed a *chip sequence*) that directly multiplies the information symbol stream. Since the chip sequence has a rate much higher than the data rate, the bandwidth is increased. The simplest form of DS/SS uses binary phase shift keying (BPSK) modulation with BPSK spreading and is illustrated in Figure 2.1. Note that this is equivalent to a standard BPSK system with a matched filter receiver with the

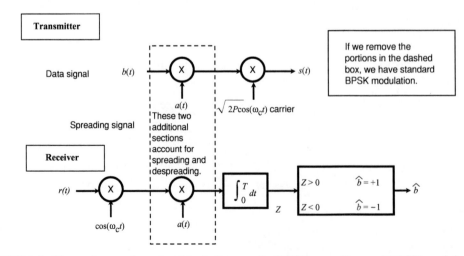

FIGURE 2.1: Transmitter and receiver block diagram for BPSK spreading and BPSK modulation.

addition of the spreading and despreading process. The receiver is equivalent to a matched filter for a DS/SS signal provided that square pulses are used. If pulse shaping is employed, the simple integrator should be replaced by a filter that is matched to the pulse shape used. If the pulse shape is incorporated at the chip level, it should also be incorporated into $a(t)$. The transmit signal can be represented by

$$s(t) = \sqrt{2P}a(t) \cos (2\pi f_c t + \theta_d (t))$$
$$= \sqrt{2P}a(t)b(t) \cos (2\pi f_c t) \quad\quad (2.1)$$

where $\theta_d(t)$ is the binary phase shift due to the information sequence, $b(t) = \sum_{i=-\infty}^{\infty} b_i p_b(t - iT_b)$ is the information signal where $b_i \in \{+1, -1\}$ represent the information bits, each bit has duration T_b, $p_b(t)$ is the unit energy pulse shape used for the information waveform (assumed to be rectangular), $a(t) = \sum_{i=-\infty}^{\infty} a_i p_c (t - iT_c)$ is the spreading signal where each symbol a_i (usually called a *chip*) has duration $T_c = T_b/N$, f_c is the center frequency of the transmit signal, P is the power of the signal, and N is the bandwidth expansion factor, sometimes also called the *spreading gain*. Example waveforms for the case of rectangular pulses are given in Figure 2.2.

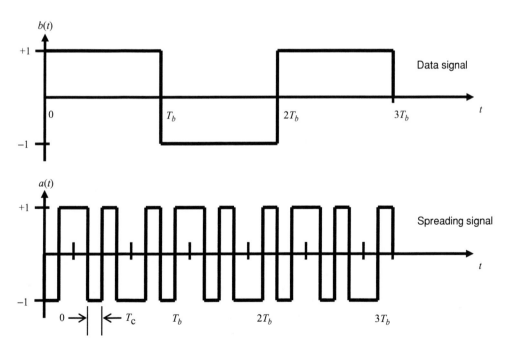

$N = T_b/T_c$ = bandwidth expansion = processing gain

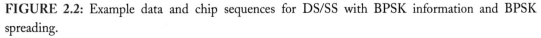

FIGURE 2.2: Example data and chip sequences for DS/SS with BPSK information and BPSK spreading.

It can be seen that the chip rate is N times that of the bit rate, resulting in a signal whose bandwidth is much larger than necessary for transmission of the information. Specifically, as we will show later, the bandwidth is commensurate with the chip rate or N times what a traditional BPSK signal bandwidth would be.

At the receiver, the opposite operations are performed. Specifically, the signal is first down-converted to baseband.[2] After down-conversion, the signal is despread and passed to a standard BPSK detector. This process can be envisioned in two ways. First, we can view the spreading/despreading operations as transparent additions to a standard BPSK transmit/receive pair. The spreading is applied after BPSK symbol creation and despreading occurs before the BPSK detector. Second, we can view DS/SS as a BPSK modulation scheme where the "pulse" is the spreading waveform. Thus, at the receiver the despreading operation can be viewed as part of a correlator version of a matched filter receiver.

At the receiver, the received signal can be modeled as

$$r(t) = s(t) + n(t)$$
$$= \sqrt{2P}a(t)b(t) \cos(2\pi f_c t) + n(t) \tag{2.2}$$

where $n(t)$ is bandpass AWGN and where the random phase offset due to propagation is assumed to be zero for simplicity. The maximum likelihood receiver then calculates the decision statistic as

$$Z = \frac{1}{T_b} \int_0^{T_b} r(t)a(t) \cos(2\pi f_c t) \, dt$$
$$= \frac{1}{T_b} \int_0^{T_b} \left(\sqrt{2P}a(t)b(t) \cos(2\pi f_c t) \right) a(t) \cos(2\pi f_c t) \, dt$$
$$+ \frac{1}{T_b} \int_0^{T_b} n(t) a(t) \cos(2\pi f_c t) \, dt$$
$$= \frac{1}{T_b} \int_0^{T_b} \left(\sqrt{2P}a^2(t)b(t) \cos^2(2\pi f_c t) \right) dt + n \tag{2.3}$$

where we have assumed perfect phase coherence, bit timing, and chip timing at the receiver and where n is a noise sample at the output of the matched filter. Now, in BPSK spreading, the spreading signal $a(t)$ can be modeled as

$$a(t) = \sum_{i=-\infty}^{\infty} a_i p_c (t - iT_c) \tag{2.4}$$

---

[2]Despreading can also be done at IF although baseband is currently more common.

where $a_i \in \{+1, -1\}$ is the spreading sequence and $p_c(t)$ is the chip pulse shape, assumed to be rectangular for this discussion. We will discuss the properties of the spreading sequence later, but for now we will assume that the chip values are random and independent. It can be readily discerned that $a^2(t) = 1$. Further, ignoring the double frequency term in (2.3), the decision statistic becomes

$$Z = \frac{\sqrt{2P}}{2} b_0 + n \tag{2.5}$$

where we have assumed that b_0 is the bit value corresponding to the interval of interest, $p_b(t)$ is a rectangular pulse of duration T_b, and n is due to AWGN and will be analyzed later. Thus, we can see that we obtain a decision variable that comprises the original bit along with a noise term, just as in standard BPSK. We will analyze the performance of this scheme shortly.

2.2.1   Power Spectral Density of Direct Sequence Spread Spectrum

The Power Spectral Density (PSD) of DS/SS depends on the modulation scheme used as well as the pulse shape used. To this point, we have assumed the use of square pulses for convenience. Based on the Wiener–Khintchine theorem, the PSD [21] of a random process is the Fourier transform of the autocorrelation function of that process. For a PAM signal of the form

$$x(t) = \sum_{i=-\infty}^{\infty} a_i p(t - i T_s) \tag{2.6}$$

where a_i are arbitrary pulse amplitudes and $p(t)$ is the pulse shape, the power spectral density can be shown to be [22]

$$S_x(f) = \frac{|P(f)|^2}{T_s} \sum_{k=-\infty}^{\infty} R_{(a,a)}[k] e^{-j2\pi f k T_s} \tag{2.7}$$

where $P(f)$ is the Fourier transform of the pulse shape, $R_{a,a}[k] = \overline{a_i a_{i+k}}$ is the autocorrelation function of the data sequence, and T_s is the symbol duration. Now, if the data is uncorrelated,[3]

$$R_{a,a}[k] = \begin{cases} \overline{a_i^2} & k = 0 \\ \overline{a_i a_{i+k}} & k \neq 0 \end{cases}$$

$$= \begin{cases} \sigma_a^2 + m_a^2 & k = 0 \\ m_a^2 & k \neq 0 \end{cases} \tag{2.8}$$

[3]Note that this is an approximation for the DS/SS spreading waveform since the spreading code is pseudo-random and periodic. For extremely long spreading codes, however, this approximation is very good.

where m_a and σ_a^2 are the mean and variance of the data amplitude sequence, respectively. Returning to the power spectral density, we have

$$
\begin{aligned}
S_x(f) &= \frac{|P(f)|^2}{T_s} \sum_{k=-\infty}^{\infty} R_{a,a}[k] e^{-j2\pi f k T_s} \\
&= \frac{|P(f)|^2}{T_s} \left(\sigma_a^2 + m_a^2 \sum_{k=-\infty}^{\infty} e^{-j2\pi f k T_s} \right) \\
&= \frac{|P(f)|^2}{T_s} \left(\sigma_a^2 + m_a^2 \sum_{k=-\infty}^{\infty} \delta\left(f - \frac{k}{T_s}\right) \right) \\
&= \frac{\sigma_a^2}{T_s} |P(f)|^2 + \frac{m_a^2}{T_s} \sum_{k=-\infty}^{\infty} \left| P\left(\frac{k}{T_s}\right) \right|^2 \delta\left(f - \frac{k}{T_s}\right)
\end{aligned}
\tag{2.9}
$$

Now for phase modulation, $m_a = 0$ and $\sigma_a^2 = 1$. Further, if square pulses are assumed, $P(f) = T_s \operatorname{sinc}(T_s f)$. Thus,

$$
S_x(f) = T_s \operatorname{sinc}^2(T_s f)
\tag{2.10}
$$

Since both the spreading waveform and the data waveform have the same format, we have the power spectral density of both. Now it remains to find the PSD of the transmitted waveform.

The complex baseband version of the transmitted signal $\tilde{s}(t)$ is an ergodic random process and the power spectral density can be found from the Fourier transform of the autocorrelation function. Since the data and the spreading sequence are independent, the autocorrelation function of the transmit signal is the product of the autocorrelation functions of the two signals. That is, since

$$
\tilde{s}(t) = \sqrt{P} a(t) b(t)
\tag{2.11}
$$

then

$$
\begin{aligned}
R_{s,s}(\tau) &= E\left\{ \tilde{s}(t) \tilde{s}^*(t+\tau) \right\} \\
&= E\left\{ a(t) b(t) a(t+\tau) b(t+\tau) \right\} \\
&= E\left\{ a(t) a(t+\tau) \right\} E\left\{ b(t) b(t+\tau) \right\} \\
&= R_{a,a}(\tau) R_{b,b}(\tau)
\end{aligned}
\tag{2.12}
$$

The power spectral density is then the Fourier transform of the autocorrelation function:

$$
\begin{aligned}
S_x(f) &= \int_{-\infty}^{\infty} S_b(\phi) S_a(f - \phi)\, d\phi \\
&= \int_{-\infty}^{\infty} T_b \operatorname{sinc}^2(\phi T_b)\, T_c \operatorname{sinc}^2([f - \phi] T_c)\, d\phi \\
&= \int_{-\infty}^{\infty} T_b \operatorname{sinc}^2(\phi T_b) \frac{T_b}{N} \operatorname{sinc}^2\left([f - \phi] \frac{T_b}{N}\right) d\phi
\end{aligned}
\tag{2.13}
$$

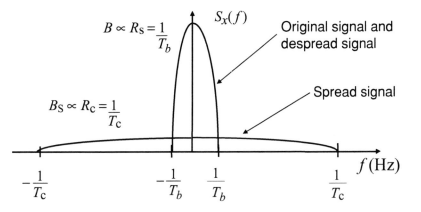

FIGURE 2.3: Illustration of PSD of original and spread signals with DS/SS.

Now, examining the last line in (2.13), we can see that if $N \gg 1$, the second term will be approximately constant over the significant values of the first term. Thus,

$$S_x(f) \approx \frac{T_b}{N} \operatorname{sinc}^2 \left(\frac{fT_b}{N} \right) \int_{-\infty}^{\infty} T_b \operatorname{sinc}^2 (\phi T_b) \, d\phi$$

$$\approx \frac{T_b}{N} \operatorname{sinc}^2 \left(\frac{fT_b}{N} \right) \qquad\qquad (2.14)$$

An illustrative sketch of the spectra (main lobe only) for the original information signal and the signal after spreading is plotted in Figure 2.3. A more concrete example is plotted in Figure 2.4 for $N = 256$. From the perspective of the spread signal, the information signal $s_b(t)$ appears to be a strong narrowband tone. From the perspective of the narrowband signal (see inset), the spread signal appears to be white noise. Further, we can see that the first-null bandwidth of the spread signal is N times that of the original information signal. Thus, we call N the *bandwidth expansion factor*, which is closely related to the processing gain and which we will discuss in Section 2.6.

The PSD of a bandpass signal can easily be found from the PSD its complex baseband representation as [22]

$$S_{bp}(f) = \frac{1}{4} \left(S_x(f - f_c) + S_x(-f - f_c) \right) \qquad\qquad (2.15)$$

Thus, the PSD of the DS/SS BPSK signal is

$$S_{bp}(f) \approx \frac{PT_c}{4} \left[\left(\frac{\sin \pi (f - f_c) T_c}{\pi (f - f_c) T_c} \right)^2 + \left(\frac{\sin \pi (f + f_c) T_c}{\pi (f + f_c) T_c} \right)^2 \right] \qquad\qquad (2.16)$$

where P is the bandpass power.

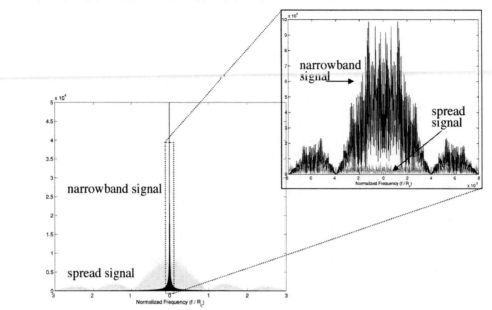

FIGURE 2.4: Comparison of original/despread signal and spread signal spectra for DS/SS with square pulses.

2.2.2   Multiple Access

The DS/SS technique can be expanded to multiple users by providing different spreading codes $a_k(t)$ to each user in the system where $a_k(t)$ is the spreading code of the kth user. A three-user example is shown in Figure 2.5. Channels are thus defined by the spreading waveform $a_k(t)$. Ideally, we would like all channels to be orthogonal. The key to making the spreading waveforms orthogonal is the spreading sequences, i.e., the series of binary values used to modulate the spreading waveform. Thus, we wish

$$R_{i,k}[n] = \frac{1}{N} \sum_{m=0}^{N-1} a_i[m]\, a_k[m+n]$$

$$= \begin{cases} 1 & i = k, n = 0 \\ 0 & i \neq k \end{cases} \qquad (2.17)$$

where $R_{i,k}[n]$ is the cross-correlation between spreading sequences a_i and a_k with a relative sequence offset n. For arbitrary n, this cannot be guaranteed. However, for $n = 0$ (i.e., synchronous codes), we can guarantee this. Specifically, we can use Walsh codes for spreading waveforms and achieve orthogonality. Walsh codes are based on Haddamard matrices, which

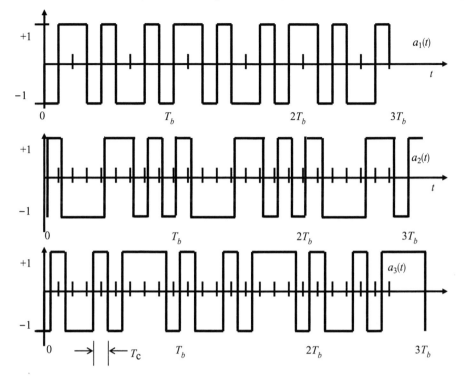

FIGURE 2.5: Illustration of CDMA based on DS/SS.

are formed as

$$H_1 = [1] \tag{2.18a}$$

$$H_{2^{i+1}} = \begin{bmatrix} H_{2^i} & H_{2^i} \\ H_{2^i} & -H_{2^i} \end{bmatrix} \tag{2.18b}$$

Walsh codes of length 2^i are then formed from the rows of the Haddamard matrix H_{2^i}. Note that the rows of the Haddamard matrix are orthogonal, and thus can be used to form 2^i orthogonal spreading codes. The length of the codes are restricted to be a power of two. Additionally, orthogonality is obtained only when the codes are aligned properly in time (i.e., synchronous). The cross-correlation properties of the codes are poor for non-synchronous alignment. Additionally, the autocorrelation properties are poor. In general, Walsh codes must be augmented with other codes to mitigate this shortcoming for synchronization purposes. The poor cross-correlation properties are demonstrated in Figure 2.6. The cross-correlation is plotted for various Walsh codes of length 64. We can see that only at zero delay (i.e., $n = 0$) do we obtain $R_{i,j}n = 0$

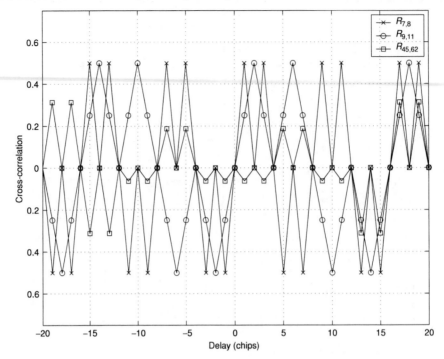

FIGURE 2.6: Example cross-correlation functions for Walsh codes of length 64.

for any i and j. Thus, Walsh codes are useful for channelization only when the codes can be guaranteed to be time synchronous at the receiver as in broadcast channels.

Walsh codes are good candidates for spreading codes only for synchronous systems; they are clearly poor candidates for asynchronous systems. However, there are several families of codes that have good properties regardless of synchronism, namely m-sequences [23], Gold codes [24], and Kasami sequences [1, 25, 26]. All three codes have excellent autocorrelation and cross-correlation properties and are easy to generate using linear feedback shift registers (LFSRs) [1].

A n-length LFSR can be in 2^n different possible states. However, if the LFSR is in the all-zeros state, it will never leave that state. Thus, the maximum possible length for an LFSR is $2^n - 1$ corresponding to the $2^n - 1$ non-zero states. One family of sequences that has length $2^n - 1$ is the maximal length sequence or the m-sequence, so named because its sequences are of maximal length. The m-sequences have many other desirable properties. One of the key properties is the two-valued periodic autocorrelation function, taking on the values

$$R_{k,k}[b] = \begin{cases} 1 & b = lN \\ -\frac{1}{N} & b \neq lN \end{cases} \tag{2.19}$$

where $N = 2^n - 1$ is the length of the sequence and l is an integer.

TABLE 2.1: Code length N and the number of m-sequences $N_p(n)$.

n	$N = 2^n - 1$	$N_p(n)$
2	3	1
3	7	2
4	15	2
5	31	6
6	63	6
7	127	18
8	255	16
9	511	48
10	1023	60

The cross-correlation between a preferred pair of m-sequences can be shown to be three-valued. More specifically,

$$R_{k,m}[b] \in \left\{ -\frac{1}{N}\left(1 + 2^{0.5(n+2)}\right), -\frac{1}{N}, \frac{1}{N}\left(2^{0.5(n+2)} - 1\right) \right\} \qquad (2.20)$$

$\forall k \neq m$ the length of the code. Unfortunately, the number of preferred pairs for a given value of N is rather limited. Thus, while m-sequences have excellent autocorrelation properties, they are not great candidates for CDMA systems because only a small number of m-sequences exist for any given sequence length $N = 2^n - 1$ as can be seen in Table 2.1 [25].

However, Gold showed in 1967 that preferred pairs of m-sequences can be combined to form sequences called Gold codes [24]. Specifically, $N + 2$ Gold codes can be created from a preferred pair of m-sequences of length N. The set of $N + 2$ sequences come from modulo-2 summing an m-sequence with N phases of the other half of a preferred pair as well as the original two sequences. The cross-correlation between any two of these Gold codes is three-valued:

$$R_{k,m}[b] \in \left\{ -\frac{1}{N}\left(1 + 2^{0.5(n+2)}\right), -\frac{1}{N}, \frac{1}{N}\left(2^{0.5(n+2)} - 1\right) \right\} \qquad (2.21)$$

for n even and

$$R_{k,m}[b] \in \left\{ -\frac{1}{N}\left(1 + 2^{0.5(n+1)}\right), -\frac{1}{N}, \frac{1}{N}\left(2^{0.5(n+1)} - 1\right) \right\} \qquad (2.22)$$

for n odd.

The benefit of Gold codes is that a large number of these codes are available for a given length N while having controlled cross-correlation properties. However, the downside is that the autocorrelation properties of Gold codes are inferior to those of m-sequences. Instead of the autocorrelation being nearly ideal with all out-of-phase values being equal to $-(1/N)$, the out-of-phase autocorrelation values can take on the same three values as the cross-correlation. Thus, the maximum absolute value of the autocorrelation function is $(1/N)\left(1 + 2^{0.5(n+2)}\right)$.

Welch showed that the maximum cross-correlation between any two sequences in a set of length N sequences of cardinality M is lower bounded [27]. Specifically, he showed that the maximum cross-correlation between two sequences is lower bounded by $\sqrt{(M-1)/(MN-1)}$ where M is the number of codes in the set. Thus, for relatively large sets, the maximum cross-correlation is greater than $\sqrt{1/N}$. From our discussion of Gold codes, we know that the maximum cross-correlation is

$$\max_{b,k,m} R_{k,m}[b]^{\text{Gold}} \approx \sqrt{\frac{2}{N}} \qquad (2.23)$$

for n odd and

$$\max_{b,k,m} R_{k,m}[b]^{\text{Gold}} \approx \sqrt{\frac{4}{N}} \qquad (2.24)$$

for n even. Thus, the maximum cross-correlation of Gold codes is higher by at least a factor of $\sqrt{2}$ than that of optimal codes. Another set of codes called Kasami sequences [1, 25, 26] can be constructed from m-sequences. These sequences have a cardinality of $2^{n/2}$ for a length of $N = 2^n - 1$ and have a three-valued cross-correlation function. Specifically, the cross-correlation function takes on values from the set $\left\{-(1/N), -(1/N)\left(2^{n/2} + 1\right), (1/N)\left(2^{n/2} - 1\right)\right\}$, which satisfies the Welch lower bound.[4] Kasami sequences are formed in a manner similar to Gold codes. We start with an m-sequence $\mathbf{a}$ of length $N = 2^n - 1$ where n is even. By decimating the sequence by $2^{n/2} + 1$, we obtain a second m-sequence $\mathbf{a}'$ of length $2^{n/2} - 1$. By adding (modulo two) $\mathbf{a}$ and $2^{n/2} - 1$ shifted versions of $\mathbf{a}'$, we obtain a set of $2^{n/2} - 1$ sequences. By also including the original sequence $\mathbf{a}$, we can ultimately obtain $M = 2^{n/2}$ total sequences. Unfortunately, while this set satisfies the Welch bound, this is not a very large set and is often called the *small* set of Kasami sequences. A larger set of Kasami sequences can be obtained, which includes Gold sequences and the small set of Kasami sequences provided that $\text{mod}(n, 4) = 2$. Again let $\mathbf{a}$ be an m-sequence of length $N = 2^n - 1$. Now, let sequences $\mathbf{a}'$ and $\mathbf{a}''$ be formed by decimating the original sequence $\mathbf{a}$ by $2^{n/2} + 1$ and $2^{(n+2)/2} + 1$. The first is a length $2^{n/2} - 1$ m-sequence, but the second is another length $2^n - 1$ m-sequence. We can form the small set of Kasami sequences by modulo-2 summing $\mathbf{a}$ with shifted versions of $\mathbf{a}'$. If we further take

---

[4]This can be readily seen by recalling that $N = 2^n - 1$.

all sequences generated by summing $\mathbf{a}$ and $\mathbf{a}''$, we obtain a set of $2^n - 1$ Gold codes. We can also obtain another $2^{n/2} - 1$ codes by modulo-2 summing $\mathbf{a}'$ and $\mathbf{a}''$. Finally, we can obtain $(2^{n/2} - 1)(2^n - 1)$ by summing all phases of $\mathbf{a}$, $\mathbf{a}'$, and $\mathbf{a}''$. Including $\mathbf{a}$ and $\mathbf{a}''$, we thus have $M = 2^{3n/2} + 2^{n/2}$ total codes. All autocorrelation and crosscorrelation values from members of this set are limited to the set $\{-1/N, -(1/N)(1 \pm 2^{n/2}), -(1/N)(1 \pm 2^{n/2+1})\}$.

As a final note, we should mention that in our discussion of spreading codes, we have assumed that the spreading waveform repeats for every bit (often termed *short codes*), and, thus the cross-correlation between user signals depends on the integration over a full sequence length. However, it is often advantageous to use long pseudo-random codes that do not repeat each bit. The performance of these codes (often called *long codes*) depends on the partial correlation properties of the codes, which are more difficult to bound.

2.3   FREQUENCY HOPPING

The goal of spread spectrum systems is to increase the dimensionality of the signal. By increasing the dimensionality, we make eavesdropping and/or jamming more difficult since there are more dimensions of the signal to consider. In commercial applications, this means that the increased dimensionality provides robustness in the presence of other signals and less interference caused to those same signals. The main method of increasing the dimensionality of the signal is to increase the signal's spectral occupancy. In Section 2.2, we discussed in detail one method of accomplishing this—DS/SS. In DS/SS, the bandwidth is increased by directly multiplying the data signal by a higher-rate pseudo-random spreading sequence. A second method of accomplishing this bandwidth expansion is through frequency hopping. In FH/SS, the carrier frequency of the data modulated sinusoidal carrier is periodically changed over some predetermined bandwidth. By "hopping" the center frequency to one of N contiguous but non-overlapping frequency bands, the overall spectrum occupancy is increased by the factor N. This hopping is typically done in a pseudo-random manner. In military applications, this makes interception and jamming more difficult. In commercial applications, it reduces the impact of a particular co-channel interferer to the frequency-hopped signal as well as the impact of the frequency-hopped signal to another system since it will be present in a particular band on average only $1/N$ of the time.

The hopping signal can be represented as

$$\eta(t) = \sum_{i=-\infty}^{\infty} p(t - i T_c) \cos(2\pi f_i t + \phi_i) \qquad (2.25)$$

where $p(t)$ is the pulse shape used for the hopping waveform (typically assumed to be a square pulse), $f_i \in \{f_1, f_2, \ldots, f_N\}$ are the N hop frequencies, T_c is the hop period also called the

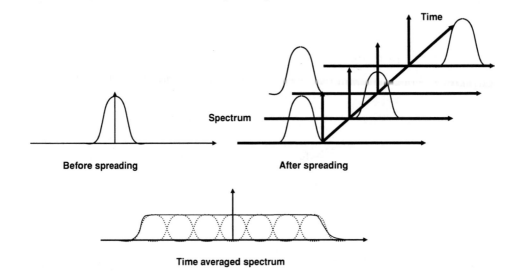

FIGURE 2.7: Illustration of spectrum spreading through frequency hopping.

chip period, and ϕ_i are the phases of each oscillator. The resulting frequency-hopped transmit signal is then

$$
\begin{aligned}
s(t) &= [s_d(t)\eta(t)]_{BPF} \\
&= \left[s_d(t) \sum_{i=-\infty}^{\infty} p(t - nT_c) \cos(2\pi f_i t + \phi_i) \right]_{BPF}
\end{aligned}
\qquad (2.26)
$$

where $s_d(t)$ is the bandpass data signal that depends on the modulation scheme employed and the bandpass filter (applied to the quantity within $[\cdot]_{BPF}$) is designed to transmit the sum frequencies only. The concept of frequency hopping is illustrated in Figure 2.7. As time advances, the signal occupies a separate frequency band as determined by the hopping sequence. On average, the power spectral density is spread over the entire band as shown. Provided each frequency band is used $1/N$ of the time, the spectrum will be similar to that seen in DS/SS systems when averaged over a sufficiently long time period.

The transmitter and receiver for a typical implementation are shown in Figures 2.8 and 2.9, respectively. As shown in the figures, any modulation scheme (with either coherent or non-coherent demodulation) can theoretically be used. As in DS/SS, the frequency hopping is ideally transparent to the data demodulator. The data modulated carrier is hopped to one of N carrier frequencies every "chip" period T_c, which may be greater than the data symbol period T_s. At the receiver, the same hopping pattern is generated such that the received signal is ideally mixed back down to the original carrier frequency. Data demodulation is then accomplished as in standard digital communications. Note that the bandwidth expansion factor is equal to

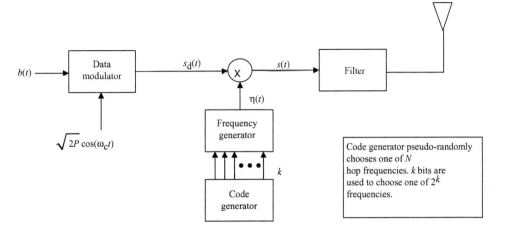

FIGURE 2.8: Typical frequency-hopping transmitter architecture.

the number of hop frequencies N. Unlike in DS/SS, the bandwidth expansion is independent of the chip period T_c. In fact, as mentioned, the chip period can be greater than the symbol period. In other words, the hopping may be slower than the symbol rate. We will discuss the consequences of this relationship later.

Although any modulation format can be used with FH/SS, coherent demodulation techniques require that the frequency hopping maintain frequency coherence each hop. This can be difficult to maintain, and thus non-coherent demodulation techniques are more commonly used with FH/SS. Specifically, M-FSK (M-ary frequency shift keying) is commonly used in conjunction with FH/SS.

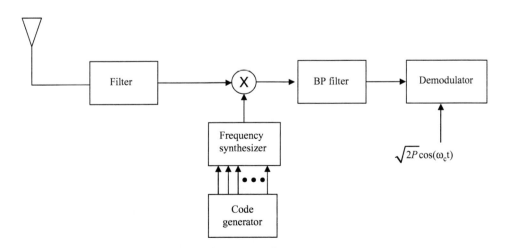

FIGURE 2.9: Typical frequency-hopping receiver architecture.

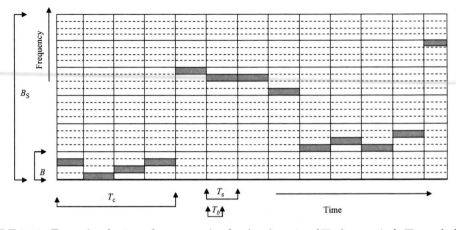

FIGURE 2.10: Example of a time–frequency plot for slow hopping (T_c : hop period, T_s: symbol period, T_b: bit period, B_s: spread bandwidth, B: symbol bandwidth).

2.3.1 Slow Versus Fast Hopping

As mentioned earlier, the hop period (also called the chip period, T_c) may be greater or less than the symbol duration. The bandwidth expansion factor is related only to the number of hop frequencies N, not the hop period. Thus, we are free to choose the hop period based on other considerations. Specifically, the hop frequency should be chosen on the basis of implementation and performance considerations. First, let us consider the case where $T_c > T_s$, or slow hopping. Additionally, let us assume that FSK modulation is used. Figure 2.10 plots an example of frequency occupancy versus time considering both the data modulation and frequency hopping. In this example, $T_c = 4T_s$, or the frequency is hopped every four symbols, $N = 6$, and 4-FSK or $M = 4$ ($T_b = T_s/2$). Further, in the figure we have defined B as the bandwidth of the M-FSK signal and B_S as the spread bandwidth. As can be seen, every T_s seconds, the frequency is changed to one of four symbols based on the data. Additionally, every T_c seconds, the center frequency of these symbols is changed on the basis of the frequency hopping pattern. At the receiver, the pseudo-random hopping is removed, leaving only the data modulation as shown in Figure 2.11.

In contrast to slow hopping, with fast frequency hopping, $T_c < T_s$, i.e., frequency hopping occurs faster than the modulation. This is depicted in Figure 2.12 where $T_c = T_s/2$, $N = 6$, and $M = 4$. In this case, coherent modulation is extremely difficult since it would require

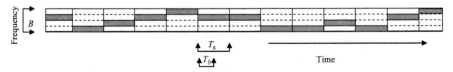

FIGURE 2.11: Example of time–frequency plot after despreading.

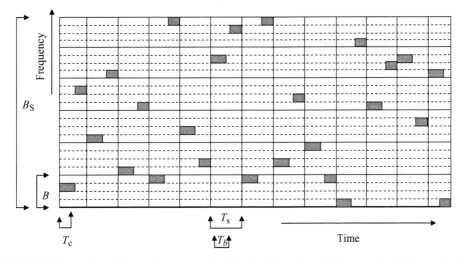

FIGURE 2.12: Example of time–frequency plot for fast hopping (T_c: hop period, T_s: symbol period, T_b: bit period, W_h: spread bandwidth, W_d: symbol bandwidth).

extremely fast carrier synchronization. Thus, non-coherent FSK is almost universally used with fast hopping. The despread or de-hopped signal is plotted in Figure 2.11, which shows that the despread data is the same as in slow hopping. Fast hopping, although more difficult to implement, offers some advantages over slow hopping. First, unlike slow hopping, fast hopping provides frequency diversity at the symbol level, which provides a substantial benefit in fading channels or versus narrowband jamming. Slow hopping can obtain these same benefits through error correction coding as we will see later, but fast hopping offers this benefit before coding, which can provide better performance, especially when punctured codes are used.

The reception of FH/SS is accomplished as shown in Figure 2.9. The despread signal $y(t)$ is obtained by multiplying the incoming signal by the hopping signal and filtering out the images:

$$
\begin{aligned}
y(t) &= [r(t)\eta(t)]_{BPF} \\
&= \left[(s(t) + n(t)) \sum_{i=-\infty}^{\infty} p(t - nT_c) \cos(2\pi f_i t + \phi_i) \right]_{BPF} \\
&= s_d(t) + n'(t)
\end{aligned}
\tag{2.27}
$$

where $n'(t)$ is the noise process after despreading and filtering.

2.3.2 Power Spectral Density of Frequency-Hopped Spread Spectrum
The power spectral density of FH/SS can be found as

$$
S(f) = S_d(f) * H(f)
\tag{2.28}
$$

where $S_d(f)$ is the power spectral density of the data modulated carrier before hopping and $H(f)$ is the power spectral density of the hopping waveform. If we define N as the number of hop frequencies, it can be shown that the PSD of the hopping waveform is [19]

$$H(f) = \frac{1}{T_c^2} \sum_{i=-\infty}^{\infty} \left| \sum_{k=1}^{N} p_k G_k \left(\frac{i}{T_c} \right) \right|^2 \delta \left(f - \frac{i}{T_c} \right) + \frac{1}{T_c} \sum_{k=1}^{N} p_k(1 - p_k) \left| G_k(f) \right|^2$$

$$- \frac{1}{T_c} \sum_{\substack{k=1 \\ m=1 \\ m \neq k \\ m > k}}^{N} \sum^{N} p_k p_m \Re \left\{ G_k(f) G_m^*(f) \right\} \qquad (2.29)$$

where $G_m(f)$ is the Fourier transform of the pulsed carrier $p(t) \cos (2\pi f_m t + \phi_m)$ defined over $0 \leq t \leq T_c$, $\Re\{x\}$ takes the real part of x, and p_m is the probability of using the mth carrier. Defining $\mathcal{F}(x(t))$ as the Fourier transform of $x(t)$, we can find $G_m(f)$ as (assuming $p(t)$ is a square pulse)

$$
\begin{aligned}
G_m(f) &= \mathcal{F} \left\{ g_m(t) \right\} \\
&= \mathcal{F} \left\{ p(t) \cos (2\pi f_m t + \phi_m) \right\} \\
&= T_c e^{-j[\pi(f - f_m)T_c - \phi_m]} \operatorname{sinc} ((f - f_m)T_c) \\
&\quad + T_c e^{-j[\pi(f + f_m)T_c + \phi_m]} \operatorname{sinc} ((f + f_m)T_c)
\end{aligned}
\qquad (2.30)
$$

If the carrier spacing is such that the spectra of $G_m(f)$ and $G_k(f)$ do not overlap for $m \neq k$ (i.e., if the hop rate $1/T_c$ is slow compared to the minimum carrier spacing) and we assume that all hop frequencies are equally likely, we obtain

$$H(f) \approx \frac{1}{T_c^2 N^2} \sum_{i=-\infty}^{\infty} \sum_{k=1}^{N} \left| G_k \left(\frac{i}{T_c} \right) \right|^2 \delta \left(f - \frac{i}{T_c} \right)$$

$$+ \frac{1}{T_c} \frac{1}{N} \left(1 - \frac{1}{N} \right) \sum_{k=1}^{N} |G_k(f)|^2 \qquad (2.31)$$

Inserting (2.30) into (2.31), the resulting PSD is

$$H(f) \approx \frac{1}{N^2} \sum_{i=-\infty}^{\infty} \sum_{k=1}^{N} \left(\operatorname{sinc}^2 (i - f_k T_c) + \operatorname{sinc}^2 (i + f_m T_c) \right) \delta \left(f - \frac{i}{T_c} \right)$$

$$+ \frac{T_c}{N} \left(1 - \frac{1}{N} \right) \sum_{k=1}^{N} \left[\operatorname{sinc}^2 ((f - f_k)T_c) + \operatorname{sinc}^2 ((f + f_k)T_c) \right] \qquad (2.32)$$

If we choose the frequency spacing to be an integer multiple of the hop rate for illustration purposes, we sample the sinc function at integer values eliminating all terms except the first:

$$H(f) \approx \frac{1}{N^2} \sum_{k=1}^{N} [\delta(f - f_k) + \delta(f + f_k)]$$

$$+ \frac{T_c}{N} \left(1 - \frac{1}{N}\right)$$

$$\cdot \sum_{k=1}^{N} \left[\text{sinc}^2\left((f - f_k)T_c\right) + \text{sinc}^2\left((f + f_k)T_c\right) \right] \qquad (2.33)$$

As an example, let us consider the power spectral density when BPSK with coherent frequency hopping is used. Now, from previous developments, we know that the PSD of BPSK is

$$S_d(f) = \frac{P T_b}{2} \left[\text{sinc}^2\left((f - f_c)T_b\right) + \text{sinc}^2\left((f + f_c)T_b\right) \right] \qquad (2.34)$$

To find the PSD of the transmit signal $S(f)$, we must convolve $H(f)$ with $S_d(f)$, resulting in [19]

$$S(f) \approx \frac{P T_b}{2N^2} \sum_{k=1}^{N} \left[\text{sinc}^2\left((f - f_k - f_c)T_b\right) + \text{sinc}^2\left((f + f_k + f_c)T_b\right) \right]$$

$$+ \left(1 - \frac{1}{N}\right) \frac{P T_b}{2N}$$

$$\cdot \sum_{k=1}^{N} \left[\text{sinc}^2\left((f - f_k - f_c)T_b\right) + \text{sinc}^2\left((f + f_k + f_c)T_b\right) \right]$$

$$= \frac{P T_b}{2N} \sum_{k=1}^{N} \left[\text{sinc}^2\left((f - f_k - f_c)T_b\right) + \text{sinc}^2\left((f + f_k + f_c)T_b\right) \right] \qquad (2.35)$$

which is an intuitively satisfying result as it says that the PSD of the frequency-hopped signal is the sum of N replicas of the information signal PSD each centered at the hopping frequencies. An example is plotted in Figure 2.13 for $R_b = 1$Mbps and hop frequencies of 11, 12, 13, and 14 MHz (i.e., four hop frequencies).

2.3.3 Multiple Access

Frequency hopping, like direct sequence, can be used for multiple access by assigning different spreading waveforms to each user. Different spreading waveforms would result in different hopping patterns as shown in Figure 2.14 and could potentially result in collisions, i.e., two signals use the same hop frequency at the same time. Ideally, we would like the hopping patterns to be orthogonal, thus avoiding collisions. This is possible only if system wide synchronization is

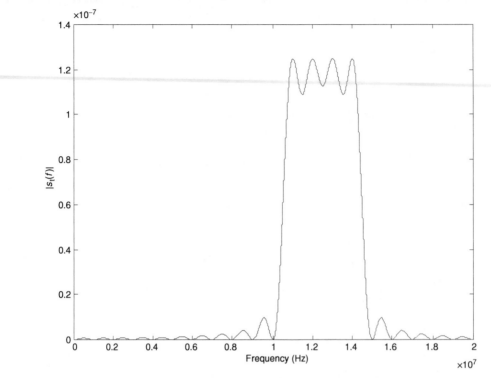

FIGURE 2.13: Example power spectrum of frequency-hopped signal with BPSK modulation ($R_b = 1\,\text{Mbps}$, $f_b = 11, 12, 13, 14\,\text{MHz}$).

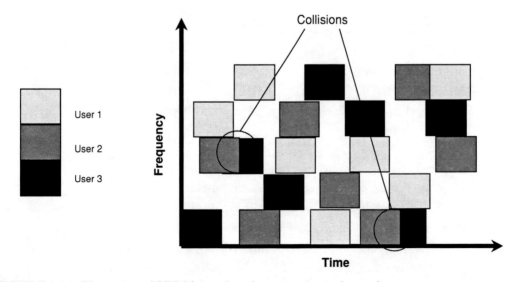

FIGURE 2.14: Illustration of CDMA based on frequency-hopped spread spectrum.

possible. As in DS-CDMA systems, synchronization is practical on the downlink of a centralized system but very difficult to maintain on the uplink. Such synchronization is nearly impossible in a decentralized system.

2.4   TIME HOPPING

As a final example of CDMA, consider a pulse-based system where transmissions are based on a series of low duty cycle pulses as in impulse-radio systems [28]. Let $s_k(t)$ be the transmit signal of the kth user. Assuming binary pulse position modulation (PPM) (similar to DS/SS), we can write the signal as

$$s_k(t) = \sum_{j=-\infty}^{\infty} \sqrt{E_p} w_{tx}\left(t - jT_f - c_j^{(k)} T_c - \delta d_{\lfloor j/N_s \rfloor}^{(k)}\right) \qquad (2.36)$$

where

- $w_{tx}(t)$ is the unit-energy transmit pulse
- E_p is the pulse energy
- N_s is the pulse repetition number, or the number of pulses used to represent one data symbol (similar to N in DS/SS)
- δ is the PPM time delay parameter
- $d_i^{(k)}$ is the ith binary data element of the kth user
- $c_j^{(k)}$ is the jth chip of user k's time-hopping sequence $c_j^{(k)} \in \{0, 1, 2, \ldots N\}$ where N is the number of time-hope slots
- T_c is the chip duration
- T_f is the frame repetition time.

Note that $\lfloor . \rfloor$ is the floor operator. An example of CDMA with time hopping is shown in Figure 2.15.

Assume that K users communicate over the same channel and that each $s_k(t)$ is transmitted through a communication channel with impulse response $h_k(t)$. Let $r_k(t) = s_k(t) * h_k(t)$ where $*$ is the convolution operator. Then, assuming asynchronous transmissions, the signal at the front end of the receiver can be written as

$$r(t) = \sum_{k=1}^{K} r_k(t - \tau_k) + n(t) \qquad (2.37)$$

where τ_k is the propagation delay of the kth user and $n(t)$ is an additive white Gaussian noise process with double-sided power spectral density $N_0/2$.

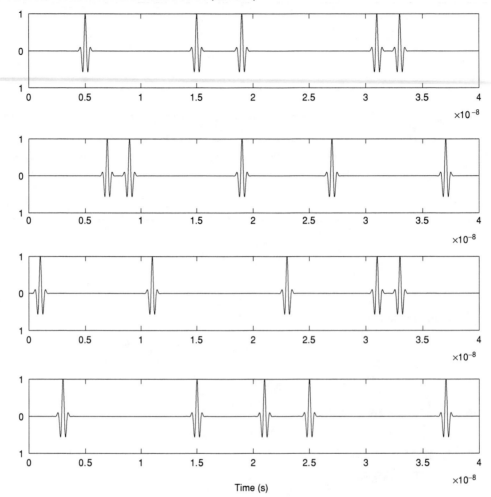

FIGURE 2.15: Example of CDMA with time hopping (four users).

Without loss of generality, we concentrate on data detection for user 1. Let $\hat{d}_0^{(1)}$ be the decision statistic for data element $d_0^{(1)}$. Then

$$\hat{d}_0^{(1)} = \sum_{j=0}^{N_s-1} \int_{\tau_1+jT_f}^{\tau_1+(j+1)T_f} r(t)v\left(t - \tau_1 - jT_f - c_j^{(1)}T_c\right) dt \qquad (2.38)$$

where

$$v(t) = w_{tx}(t) - w_{tx}(t - \delta) \qquad (2.39)$$

Decisions are based on the sign of $\hat{d}_0^{(1)}$. Note that interference averaging and fading diversity are obtained through the summation of N_s pulses per bit, similar to DS/SS. Also, the receiver structure described is optimal only in the absence of multiple access interference. Time hopping can be thought of as DS/SS with tertiary (three-valued) spreading codes where the values of the spreading code exist on $\{-1, 0, +1\}$. The performance, however, is analogous to frequency hopping since collisions occur when two users hop to the same portion of the transmit frame during a specific frame interval.

2.5   LINK PERFORMANCE OF DIRECT SEQUENCE SPREAD SPECTRUM IN CODE DIVISION MULTIPLE ACCESS

Now that we have described the main types of CDMA, let us now turn to the performance of CDMA systems. This discussion will start with DS-CDMA—specifically, with the performance of a single DS/SS link in AWGN and Rayleigh fading channels. While DS/SS provides no benefit over narrowband BPSK in AWGN, it can provide substantial benefits in fading channels. We will then discuss the performance of the DS/SS link in the presence of multiple access interference (MAI), i.e., the performance of multiple simultaneous DS/SS links.

2.5.1   Additive White Gaussian Noise

The complex baseband version of a received DS/SS signal can be represented as

$$\tilde{r}(t) = \sqrt{P}a(t)b(t)\gamma(t) + \tilde{n}(t) \tag{2.40}$$

where $a(t)$ is the spreading waveform assumed to have unit power, $b(t)$ is the modulation waveform also assumed to have unit *average* power, $\gamma(t)$ is the time-varying channel assuming that no frequency selective fading occurs, and $\tilde{n}(t)$ is AWGN modeled as a baseband complex Gaussian random process with variance σ_n^2. The maximum SNR receiver correlates the received signal with the complex conjugate[5] of the spreading waveform and pulse shape of the transmitted signal. Assuming square pulses and BPSK modulation, we have

$$Z_i = \frac{1}{T} \int_{(i-1)T}^{iT} \tilde{r}(t)a^*(t)\,dt \tag{2.41}$$

We can readily show that in an AWGN channel where $\gamma(t) = 1$ the output of the integration can be represented as

$$Z_i = \sqrt{P}b_i + n_i \tag{2.42}$$

Now the variance of n_i is equal to $(1/T^2)E_a\sigma_n^2$ where E_a is the energy in the spreading waveform over the symbol period. Since the spreading waveform has unit power, $E_a = T$. Further, since we are using a matched filter [22], we know that $\sigma_n^2 = N_0/2$. Thus, the variance at the output

---

[5]If the spreading waveform is complex, the complex conjugate of the spreading waveform is needed for the despreading process.

of the matched filter conditioned on b_i is

$$\sigma_Z^2 = \frac{1}{T^2} E_a \sigma_n^2$$

$$= \frac{1}{T^2} T \frac{N_0}{2}$$

$$= \frac{N_0}{2T} \qquad (2.43)$$

The performance of the matched filter using optimal detection is

$$P_e = Q\left(\frac{\overline{Z}}{\sigma_Z}\right)$$

$$= Q\left(\sqrt{\frac{2PT}{N_0}}\right)$$

$$= Q\left(\sqrt{\frac{2E_b}{N_0}}\right) \qquad (2.44)$$

where $Q(x) = \frac{1}{\sqrt{2\pi}} \int_x^\infty e^{\left(-\frac{u^2}{2}\right)} du$ is the standard Q-function which is identical to BPSK without spreading and is plotted in Figure 2.16. We can thus conclude that there is no advantage of DS/SS in AWGN channels when no MAI is present.

2.5.2   Multipath Fading Channels

One type of channel that is well suited to spread spectrum is the multipath fading channel. By modeling the channel as a linear system [7], the complex baseband version of the received signal $\tilde{r}(t)$ can be modeled as the convolution of the channel impulse response with the transmit signal

$$\tilde{r}(t) = s(t) \otimes h(t, \tau) + n(t) \qquad (2.45)$$

where $s(t)$ is the transmit waveform, $n(t)$ is AWGN, and $h(t, \tau)$ is the time-varying impulse response of the channel, which is typically modeled as a series of impulses [7]. The complex baseband channel can be represented as

$$h(t, \tau) = \sum_{i=1}^{L} \alpha_i(t) e^{j(2\pi f_i t + \theta_i)} \delta(\tau - \tau_i) \qquad (2.46)$$

where α_i, f_i, θ_i, and τ_i are the amplitude, Doppler shift, phase shift, and delay associated with the ith path. Note that the amplitudes of the paths are typically assumed to be normalized such that the channel has unit average gain.

Fading results when two multipath components arrive with a relative time offset that is much smaller than a symbol interval (i.e., non-resolvable) but large relative to a single cycle of the carrier frequency. For example, if the symbol rate is 1Mbps and the carrier frequency

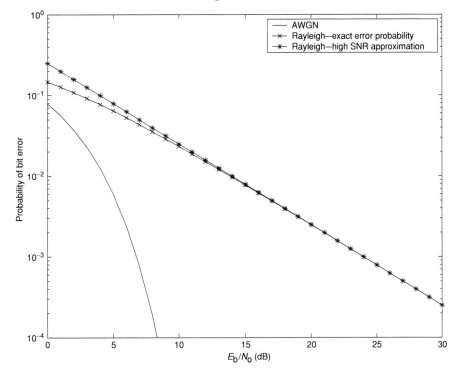

FIGURE 2.16: Performance of narrowband BPSK in AWGN and flat Rayleigh fading (DS/SS provides the same performance in an AWGN channel).

is 2GHz, a relative time offset between two multipath components of 1.25ns is insignificant relative to the symbol duration of 1 μs but results in a phase difference of 7.9 radians. When multipath components combine non-coherently, they have the potential of annihilating each other. Provided that the differences between the delays are very small compared to the symbol duration (this is called a *narrowband assumption*), all the paths appear to arrive simultaneously and we can model the channel impulse response as

$$h(t, \tau) = \sum_{i=1}^{L} \alpha_i(t) e^{j(2\pi f_i t + \theta_i)} \delta(\tau - \tau_o) = \gamma(t) \qquad (2.47)$$

The resulting received complex baseband signal can then be modeled as

$$\tilde{r}(t) = \gamma(t) s(t) \qquad (2.48)$$

where $\gamma(t)$ is a complex Gaussian random process that represents multiplicative distortion. This is termed *flat fading* because the spectrum of $h(t, \tau)$ with respect to τ has a bandwidth that is large relative to the bandwidth of $s(t)$ and thus has a frequency response that is flat over the band of $s(t)$.

The performance of a spread spectrum signal in flat Rayleigh fading can be determined similar to the AWGN case with the exception that due to fading we now must integrate the probability of error over the SNR distribution. The decision statistic for BPSK modulation can be written as the output of the matched filter, which is also a correlator output:

$$
\begin{aligned}
Z &= \frac{1}{T} \int_0^T \tilde{r}(t) a^*(t) \, dt \\
&= \frac{1}{T} \int_0^T \left(\sqrt{P} b(t) a(t) \gamma(t) + n(t) \right) a^*(t) \, dt \\
&= \frac{\sqrt{P} \gamma}{T} \int_0^T b(t) a(t) a^*(t) \, dt + \frac{1}{T} \int_0^T n(t) a^*(t) \, dt \\
&= \sqrt{P} \gamma b + n
\end{aligned}
\tag{2.49}
$$

where we have assumed that the channel $\gamma(t)$ remains constant over a symbol interval. Since the baseband channel is complex, we must remove the channel-induced phase modulation to perform phase detection. Multiplying the matched filter output by the complex conjugate of the channel gain γ^* and assuming BPSK modulation, the probability of error conditioned on γ is then

$$
P_e \mid \gamma = Q \left(\sqrt{\frac{2E_b}{N_0} |\gamma|^2} \right)
\tag{2.50}
$$

Since γ is a complex Gaussian random variable, $|\gamma|^2$ is a Chi-square random variable with two degrees of freedom. The average probability of error is the conditional probability of error averaged over the distribution of $|\gamma|^2$. The probability density function (pdf) of $\beta = \frac{E_b}{N_0} |\gamma|^2$ is

$$
p(\beta) = \frac{1}{\overline{\beta}} e^{-\frac{\beta}{\overline{\beta}}}
\tag{2.51}
$$

where $\overline{\beta} = \frac{E_b}{N_0} \overline{|\gamma|^2}$ is the average SNR at the output of the matched filter. The average probability of error is then

$$
\begin{aligned}
P_e &= \int_0^\infty p(\beta) Q\left(\sqrt{2\beta}\right) d\beta \\
&= \int_0^\infty \frac{1}{\overline{\beta}} e^{-\frac{\beta}{\overline{\beta}}} Q\left(\sqrt{2\beta}\right) d\beta \\
&= \frac{1}{2} \left(1 - \sqrt{\frac{\overline{\beta}}{1 + \overline{\beta}}} \right)
\end{aligned}
\tag{2.52}
$$

If $\bar{\beta} \gg 1$,

$$
\begin{aligned}
P_e &= \frac{1}{2}\left(1 - \frac{1}{\sqrt{1 + \frac{1}{\bar{\beta}}}}\right) \\
&\cong \frac{1}{2}\left(1 - \left(1 - \frac{1}{2\bar{\beta}}\right)\right) \\
&= \frac{1}{4\bar{\beta}}
\end{aligned}
\tag{2.53}
$$

The probability of error and the high SNR approximation are shown in Figure 2.16. As can be seen, the performance is dramatically degraded as compared to the AWGN case.

As we can see from Figure 2.16, multipath is a serious impairment to wireless communications. However, the previous analysis assumes flat Rayleigh fading (i.e., the entire frequency band of the signal fades coherently). The use of spread spectrum allows for substantial improvement in performance, one of the greatest benefits of spread spectrum communication. Specifically, the preceding analysis assumed that flat Rayleigh fading is experienced. By spreading the bandwidth well beyond the data bandwidth, we increase the likelihood that frequency selective fading will occur. Specifically, at high chip rates, the chip duration T_c is smaller than the relative multipath delay values. While this has a negative effect in traditional communication systems, it is beneficial in spread spectrum systems. This is due to the good autocorrelation properties of the spreading codes used. Further, the multipath components, in addition to exhibiting low cross-correlation, can be harnessed to provide diversity performance. The receiver structure that accomplishes this is called a *Rake receiver* and is shown in Figure 2.17.

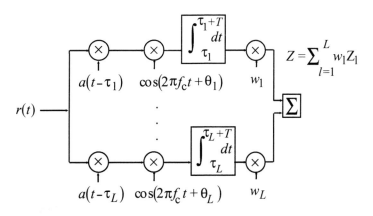

FIGURE 2.17: Block diagram of a Rake receiver.

The performance of the Rake receiver in a frequency selective channel can be found by again examining the decision statistic for BPSK. First, let us assume that the received signal comprises L resolvable[6] multipath components:

$$r(t) = \sum_{i=1}^{L} \gamma_i \sqrt{P} b(t - \tau_i) a(t - \tau_i) + n(t) \qquad (2.54)$$

where γ_i is the ith resolvable complex multipath gain and τ_i is the delay of the ith resolvable multipath component. Note that at present we have ignored the time-varying nature of the channel. The received signal is passed through L Rake correlators with each tuned to a different multipath delay as shown in Figure 2.17. The decision statistic results from the coherent combination of the L correlator outputs:

$$Z = \sum_{i=1}^{L} \gamma_i^* Z_i \qquad (2.55)$$

where $()^*$ is the complex conjugate and

$$
\begin{aligned}
Z_i &= \frac{1}{T} \int_0^T r(t) a^*(t - \tau_i) \, dt \\
&= \frac{1}{T} \int_0^T \left(\sum_{k=1}^{L} \gamma_k \sqrt{P} b(t - \tau_k) a(t - \tau_k) + n(t) \right) a^*(t - \tau_i) \, dt \\
&= \sum_{k=1}^{L} \frac{\sqrt{P}\gamma_k}{T} \int_0^T b(t - \tau_k) a(t - \tau_k) a^*(t - \tau_i) \, dt + \frac{1}{T} \int_0^T n(t) a^*(t - \tau_i) \, dt \\
&= \sqrt{P}\gamma_i b + n_i \qquad (2.56)
\end{aligned}
$$

where we have assumed that $b(t)$ is constant over the symbol interval and

$$\int_0^T a(t - \tau_k) a^*(t - \tau_i) = \begin{cases} T & i = k \\ 0 & i \neq k \end{cases} \qquad (2.57)$$

The second assumption is a crucial one. It is this that allows spread spectrum to avoid the deleterious effects of multipath fading. The assumption is justified because the autocorrelation properties of the spreading code used are nearly a delta function. If the autocorrelation function does not obey this assumption, substantial multipath interference could result.

Now, if we assume that each path exhibits Rayleigh fading, the decision statistic is the sum of L Chi-square random variables and is thus also a Chi-square random variable with $2L$

---

[6]By "resolvable" we mean that the paths are separated in time by more than T_c and thus can be separated by a matched filter receiver.

degrees of freedom. In other words, for a fixed set of channel gains, the performance is

$$P_e \mid \gamma_i = Q\left(\sqrt{\frac{2E_b}{N_0} \sum_{i=1}^{L} |\gamma_i|^2}\right) \tag{2.58}$$

and thus,

$$P_e = \int_0^\infty p(\beta) Q\left(\sqrt{2\beta}\right) d\beta \tag{2.59}$$

where β is now a central Chi-square random variable with $2L$ degrees of freedom. Assuming that all branches have equal power:

$$p(\beta) = \frac{1}{(L-1)!\overline{\beta}^L} \beta^{L-1} e^{-\frac{\beta}{\overline{\beta}}} \tag{2.60}$$

where $\overline{\beta}$ is the average SNR per path. The probability of error is then

$$\begin{aligned}
P_e &= \int_0^\infty p(\beta) Q\left(\sqrt{2\beta}\right) d\beta \\
&= \int_0^\infty \frac{1}{(L-1)!\overline{\beta}^L} \beta^{L-1} e^{-\frac{\beta}{\overline{\beta}}} Q\left(\sqrt{2\beta}\right) d\beta \\
&= \left[\frac{1}{2}\left(1 - \sqrt{\frac{\overline{\beta}}{1+\overline{\beta}}}\right)\right]^L \sum_{i=0}^{L-1} \binom{L-1+k}{k} \left[\frac{1}{2}\left(1 + \sqrt{\frac{\overline{\beta}}{1+\overline{\beta}}}\right)\right]^k
\end{aligned} \tag{2.61}$$

The resulting performance is shown in Figure 2.18 for $L = 1, 2, 4, 8$. We can see that employing a Rake receiver with spread spectrum greatly mitigates the harmful impact of Rayleigh fading. With eight equal strength paths, the performance is nearly identical to an AWGN channel.

The previous plot assumes that each path has equal power. However, this is atypical in practice. Figures 2.19–2.21 show the impact of unequal paths. Specifically, in Figure 2.19, the power of the second path in the two-path case is 3dB lower than that of the first path. In the four-path case, the four paths powers are 0, −3, −6, and −9dB relative to the first path. It can be seen that for a given number of multipath components, unequal strength causes a reduction in equivalent energy (as seen by the leftward shift of the curves) but does not reduce the diversity gain (as represented by the slope of the curves). In Figure 2.20, the same plot is given but the additional paths are even weaker. For the two-path case, the second path is 6dB weaker than the first, whereas in the four-path case, the three diversity paths are 6, 9, and 12dB lower than that of the first path. Figure 2.21 shows a more complete plot of the two-path case. It can be seen that, even with a relatively weak second path, decent diversity gains are provided. Thus, a

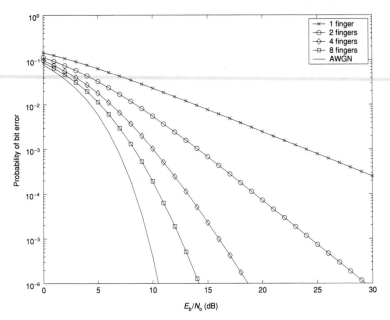

FIGURE 2.18: Bit error rate of BPSK in frequency selective Rayleigh fading with a Rake receiver. (Note that the number of resolvable multipath components equals the number of fingers.)

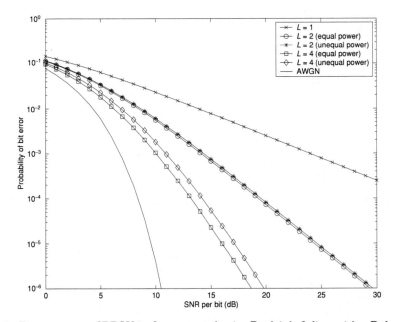

FIGURE 2.19: Bit error rate of BPSK in frequency selective Rayleigh fading with a Rake receiver with unequal path energies (second path 3dB below main for two-path case; paths 3, 6, and 9dB below the main path for the four-path case).

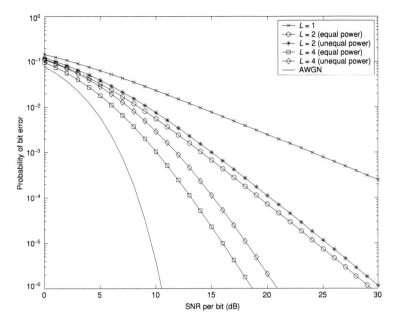

FIGURE 2.20: Bit error rate of BPSK in frequency selective Rayleigh fading with a Rake receiver with unequal path energies (second path 6dB below main for two-path case; paths 6, 9, and 12dB below the main path for the four-path case).

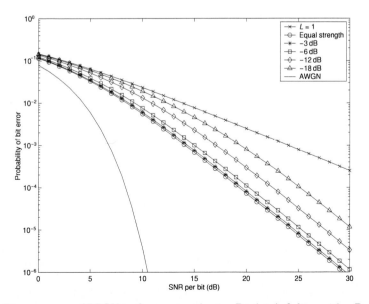

FIGURE 2.21: Bit error rate of BPSK in frequency selective Rayleigh fading with a Rake receiver with two multipath components and unequal path energies (various power levels for second path).

Rake receiver can provide substantial improvement in performance even if the environment is not rich in multipath.

2.5.3   Impact of Bandwidth

Increasing the bandwidth of a signal while keeping the data rate constant impacts the performance of the communications signals in three distinct ways. As discussed in the previous section, increasing bandwidth increases the number of resolvable multipath components. Each of these components contain a fraction of the overall energy and are independent, leading to the possibility of diversity. Thus, a receiver with a single correlator sees a degradation in performance since it captures only a portion of the energy. However, a receiver with multiple correlators (Rake fingers) can capture all the energy and coherently combine the resolvable paths to improve the performance through diversity. In most systems, there is also a third effect that occurs. Because each resolvable component is made up of fewer multipath components as the bandwidth increases, the fading per finger is actually less severe.

In other words, while we assumed in the last section that each Rake finger was subject to Rayleigh fading, this is usually found to be a pessimistic assumption, especially as the bandwidth grows. In this section, we will examine measurements taken in an indoor office environment at Virginia Tech that illustrate all three effects [29]. Measurements were taken with a sliding correlator system to examine the statistics of the received signal at several bandwidths: a continuous wave sinusoid (CW) and direct sequence chip rates of 25, 100, 225, 400, and 500MHz. The results presented here are averaged over 10,000 measurements in non-line-of-sight (NLOS) conditions with transmit distances from 2 to 20m. (See Hibbard's work [29] for more details.) In Figure 2.22, we plot the distribution of the amplitude of the strongest resolvable multipath component normalized to unit average energy. Each distribution is a Nakagami-m distribution with different m factors. The CW waveform experiences typical NLOS fading with the amplitude following a Rayleigh ($m = 1$) distribution. As the bandwidth increases, we see that the fading on the first component becomes less severe, with the most dramatic changes occurring when the bandwidth increases from 100 to 225MHz. The change in fading severity is small as the bandwidth changes from 225 to 500MHz. This can be observed in the associated Nakagami-m parameters, which were found to be $m = \{1, 1.4, 1.7, 4.9, 5.4, 5.5\}$ for a CW tone, and DS/SS signals with chip rates of 25MHz, 100MHz, 225MHz, 400MHz, and 500MHz respectively.

The second effect observed as the bandwidth increases is the reduction in the energy captured in each correlator (finger). This is illustrated in Figure 2.23, which plots the cumulative energy capture in terms of the percent of total available energy as the number of correlators (fingers) increases. If only a single correlator is used, 88% of the energy is captured when the a chip rate (bandwidth) is 25MHz. However, at 500MHz, a single correlator captures only 12%

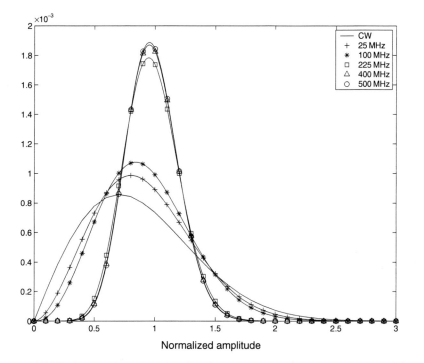

FIGURE 2.22: PDF of normalized amplitude of primary path (example from NLOS indoor office environment).

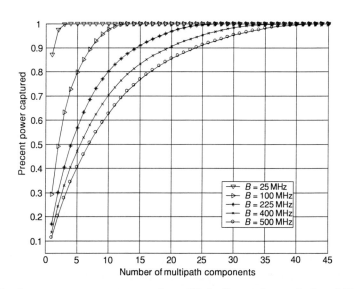

FIGURE 2.23: Total energy capture versus number of Rake fingers (example from NLOS indoor office environment).

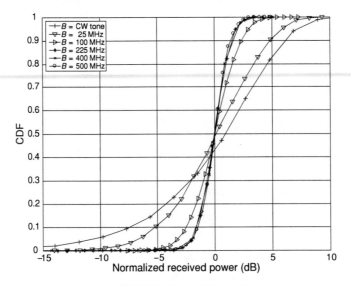

FIGURE 2.24: CDF of total energy in NLOS indoor office environment.

of the available energy. Viewed another way, if we desire to capture at least 90% of the available energy, we must use two correlators when the chip rate is 500. However, when the bandwidth is 500MHz, we must use approximately 25 Rake fingers.

Finally, if we can combine all the energy using an unlimited number of Rake fingers to achieve full energy capture, we can reduce the fading even more dramatically, as shown in Figure 2.24. This plot presents the empirical CDF of the total received energy (in decibels) for different bandwidths assuming that all the energy is captured. The severity of the fading is dramatically mitigated as the bandwidth of the signal is increased. This can be seen by the increase in slope as the bandwidth increases. The resulting impact on performance is also dramatic. Using the empirical distribution of the matched filter output for each bandwidth in (2.59) results in the bit error rate (BER) plots of Figure 2.25. At 1% BER, the highest bandwidth examined has a performance that is 8dB better than a narrowband signal due to diversity. However, this ignores the requirement for additional channel estimation, which can be substantial.

2.6   MULTIPLE ACCESS PERFORMANCE OF DIRECT SEQUENCE CODE DIVISION MULTIPLE ACCESS

The previous section examined the performance of a single DS/SS link in various environments. However, we are primarily interested in the performance of CDMA links. Thus, we must examine the performance in the presence of MAI, i.e., in a multiuser environment. In this section, we examine the performance of multiuser systems using the widely employed Gaussian approximation.

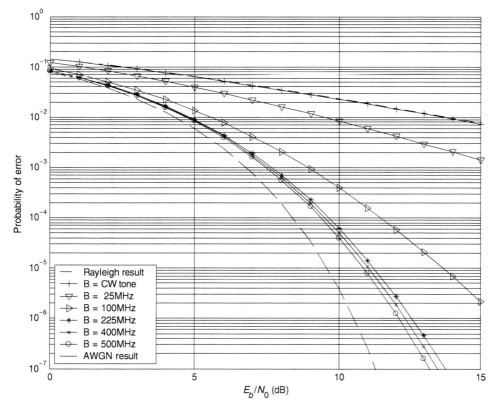

FIGURE 2.25: Performance of BPSK in measured NLOS channels with various bandwidths (assumes full energy capture [29]).

2.6.1   Gaussian Approximation

The Gaussian approximation (GA) is a method to approximate the BER of a direct sequence CDMA system by modeling the decision statistic used for symbol estimation as a Gaussian random variable. This means we must assume that MAI after despreading is well modeled as a Gaussian random variable. In a conventional system, the decision statistic, denoted by Z, is simply the output of a filter matched to the spreading code of the desired user. In a multiuser environment, the received signal can be written as

$$r(t) = \sum_{k=o}^{K-1} \sqrt{2P_k} a_k(t) b_k(t) \, \cos(\omega_c t + \phi_k) + n(t) \qquad (2.62)$$

The decision statistic is again the output of a matched filter. Analogous to (2.49), this becomes

$$Z_{k,i} = \int_{iT+\tau_k}^{(i+1)T+\tau_k} r(t) a_k(t - \tau_k) \cos(\omega_c t + \phi_k) \, dt \qquad (2.63)$$

Since there are exactly N bits in this sequence, A and B are related by $A + B = N - 1$. Assuming that the spreading sequence of user k is generated randomly (a safe assumption for long pseudo-random spreading codes), then B is also a binomial random variable with probability mass function

$$p_B(j) = \binom{N-1}{j} 2^{1-N}, \quad j = 0, \ldots, N-1 \tag{2.74}$$

The standard Gaussian approximation assumes that the variables X_j, Y_j, U_j, and V_j are all independent random variables. However, as we can see, X_j and Y_j are both dependent on the random variable B. Thus, they are not independent nor are the individual $I_{j,k}$ terms in the summation since B is a property of user k and thus the same for all $I_{j,k}$. The terms are, however, independent conditioned on B, which will be explored more later. Continuing with the standard Gaussian approximation, we can easily see that $\mathbf{E}[W_j] = 0$ since each term in (2.71) is zero mean, leading to the conclusion that $\mathbf{E}[Z_{k,i} \mid b_{k,i}] = \sqrt{(P_k/2)} T_b b_{k,i}$. Due to the assumed independence of $I_{j,k}$, the variance of the summation can be reduced to the summation of variances, i.e.,

$$\mathbf{E}\left[\left(\sum_{j=1, j\neq k}^{K} I_{j,k} \right)^2 \right] = \mathbf{E}\left[\sum_{j=1, j\neq k}^{K} (I_{j,k})^2 \right] \tag{2.75}$$

Further, assuming the independence of the terms in (2.71), we can show that

$$\mathbf{E}[W_j^2] = \mathbf{E}\left[X_j^2 + \left(1 - \frac{2\Delta_j}{T_c}\right)^2 Y_j^2 + \left(1 - \frac{\Delta_j}{T_c}\right)^2 U_j^2 + \left(\frac{\Delta_j}{T_c}\right)^2 V_j^2 \right] \tag{2.76}$$

Since U_j and V_j are binomial random variables, $\mathbf{E}[U_j^2] = \mathbf{E}[V_j^2] = 1$.

Additionally, it can be shown that given B, $\mathbf{E}[X_j^2] = N - B - 1$ and $\mathbf{E}[Y_j^2] = B$. Finally, since Δ_k is uniform on $[0, T_c)$, $\mathbf{E}[\Delta_k] = T_c/2$ and $\mathbf{E}[\Delta_k^2] = T_c^2/3$. Using these values in (2.76) results in

$$\mathbf{E}[W_j^2 \mid B] = (N - B + 1) + \frac{1}{3}B + \frac{2}{3} \tag{2.77}$$

Strictly, we should average the performance over the distribution of B. However, the standard Gaussian approximation assumes that B takes on its expected value, $\mathbf{E}[B] = (N-1)/2$. Substituting this value into (2.77) and subsequently into (2.70) leads to

$$\mathbf{E}\left[(I_{j,k})^2\right] = T_c^2 \frac{P_j}{2} \mathbf{E}[\cos^2(\phi_j)] \mathbf{E}[W_j^2] \tag{2.78a}$$

$$= T_c^2 \frac{P_j}{4} \left(\frac{3N - 2((N-1)/2) - 1}{3} \right) \tag{2.78b}$$

$$= \frac{T_c^2 P_j N}{6} \tag{2.78c}$$

The variance of the interference term is then

$$E\left[\left(\sum_{j=1, j\neq k}^{K} I_{j,k}\right)^2\right] = \frac{T_c^2 N}{6}\sum_{j\neq k} P_j \qquad (2.79)$$

giving an overall variance of the decision statistic as

$$\text{var}[Z_{k,i} \mid b_{k,i}] = \frac{N_0 T_b}{4} + \frac{T_b^2}{6N}\sum_{j\neq k} P_j \qquad (2.80)$$

where we have used the fact that $T_b = NT_c$. If this expression is used in (2.65), we arrive at the standard Gaussian approximation for P_e for user k:

$$P_e = Q\left(\sqrt{\left(\frac{N_0}{2T_b P_k} + \frac{1}{3N}\sum_{j\neq k}\frac{P_j}{P_k}\right)^{-1}}\right) \qquad (2.81)$$

There are several aspects of CDMA systems that can be inferred from (2.81). First, if $K = 1$, i.e., there is no MAI, then (2.81) simplifies to

$$P_e = Q\left(\sqrt{\frac{2E_b}{N_0}}\right) \qquad (2.82)$$

which is the performance of standard BPSK in an AWGN channel. Additionally, we can see that for a given number of users K increasing E_b/N_0 beyond a certain level does not improve performance. In other words, the system is interference-limited. This can be seen in Figure 2.27. In this plot, the average probability of bit error is plotted versus E_b/N_0 for various numbers of users K, $N = 128$, and perfect power control (i.e., $P_1 = P_2 = P_i \ \forall i$). We can see that performance saturates at high values of E_b/N_0 due to the interference limit. This lower bound on performance can easily be obtained from (2.81) by letting $E_b/N_0 \to \infty$ and is a basic consequence of MAI. Additionally, we can see that the spreading gain N has a large impact on performance. Increasing the spreading gain increases the system capacity. This can be seen from Figure 2.28 where we plot the average probability of bit error versus the number of users in the system for various values of N. Note that $E_b/N_0 = 20$dB in the plot. If a particular probability of error is required, increasing N increases capacity. For example, let us assume that a probability of error of 10^{-4} is required. For a spreading gain of $N = 16$, the system can support approximately $K = 4$ users. However, if $N = 128$, the system can support approximately $K = 26$ users. Of course, this increase in capacity comes at a price. By increasing the spreading gain for a fixed data rate, the overall bandwidth requirements increase. In fact, in the previous example, an increase in the spreading gain (and thus bandwidth) by a factor of 8 resulted in a capacity increase of only a factor of 6.5. This is a general trend in a single-cell CDMA scenario. However, there are several other factors that we have not considered that arise in the multiple cell scenario

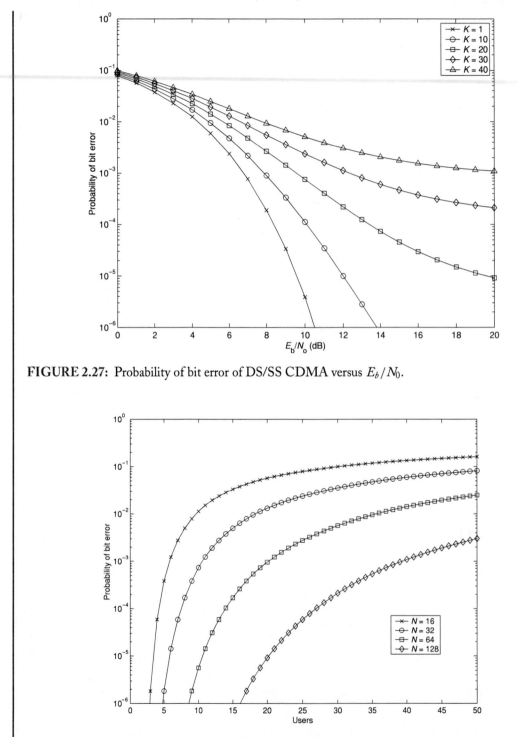

FIGURE 2.27: Probability of bit error of DS/SS CDMA versus E_b/N_0.

FIGURE 2.28: Probability of bit error of DS/SS CDMA versus the number of users.

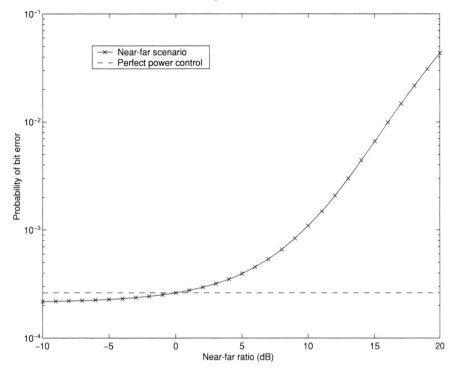

FIGURE 2.29: Impact of near-far disparity in a DS-CDMA system. The near-far ratio is defined as the ratio of the received power of the uncontrolled signal (P_U) to the received power of the controlled users (P). (In dB we have Near-Far Ratio (dB) $= 10\log_{10}(P_U) - 10\log_{10}(P)$).

and make the capacity increases much more attractive. We will discuss this in more detail in Chapter 3.

Another aspect of DS-CDMA systems that can be inferred from (2.81) is what is termed the near-far effect, which occurs whenever the received power levels of two users are substantially different. For example, consider a case where $N = 128$, $E_b/N_0 = 20$dB, and $K = 30$. If all users are received at the same power level (this situation is termed perfect power control), we can determine from (2.81) that all users obtain an average bit error rate of approximately $2.6e - 4$. However, if a signal is received from one of the thirty users with a power level that is different from the other users, not only will that user's performance change, but so will the performance of the other users. In Figure 2.29, we plot the probability of bit error for the power controlled users as an uncontrolled user's power varies from 10dB below that of the other users to 20dB above that of the other users will be affected. We can see that when it is weaker than the other signals (a near-far ratio of -10dB), it has little impact on their performance. However, as the received power of the uncontrolled user grows relative to the power-controlled users, their performance degrades rapidly. In fact, if the received signal strength is 20dB greater than that of the other signals, those signals will experience a degradation in BER of over two orders of magnitude. This necessitates strict power control in CDMA systems. Power control counteracts the differences in path loss, shadowing, and fading in the signals of various users, which can easily be 30–40dB, and thus mitigates the near-far problem.

2.6.2   Improved Gaussian Approximation

The previous discussion assumed that the interference in DS-CDMA can be approximated by a Gaussian random variable. Unfortunately, this can seriously underestimate the BER for low numbers of users [32]. A more accurate approximation considers the variance of the MAI to be a random variable with a distribution that depends on the distribution of U, V, and B [32]. Unfortunately, this distribution is difficult to compute, making the BER calculation computationally expensive. Holtzman provided a simplified expression based on the Stirling formula [33], and Liberti extended this to the case of random interference powers [7, 34]. Fortunately, most practical cases do not require this improved Gaussian approximation (IGA) because the Gaussian approximation is inaccurate only when the number of users is low, the desired BER is very low, or E_b/N_0 is very high. Most of these scenarios are not common in CDMA systems. Even for random interference powers, the IGA is unneeded because when the desired signal also varies, the power variation of the signal of interest tends to dominate performance. Thus, a reasonable approximation for fading channels is to integrate (2.81) over the distribution of the desired signal's power, P_k:

$$P_e = \int_0^\infty Q\left(\sqrt{\left(\frac{N_0}{2T_b p} + \frac{1}{3N}\sum_{j\neq k}\frac{P_j}{p}\right)^{-1}}\right) f_P(p)\, dp \qquad (2.83)$$

where $f_P(p)$ is the distribution of the desired signal's power. As an example, consider a system where all amplitudes follow independent Rayleigh distributions with equal average power and $\overline{E_b/N_0} = 20$dB. Figure 2.30 plots the BER performance versus the number of users for $N = 31$ from simulation and using the standard Gaussian approximation (labeled "theory") from (2.83). We can see that the standard Gaussian approximation is very accurate. (For more information about the IGA, please refer to [7, 32–34].)

2.7   LINK PERFORMANCE OF FREQUENCY-HOPPED SPREAD SPECTRUM

As with DS/SS, frequency-hopped spread spectrum provides unique benefits in terms of the link performance. Before we describe the performance in multiple access scenarios, it is important to examine these benefits. The most common modulation scheme with FH/SS, as discussed earlier,

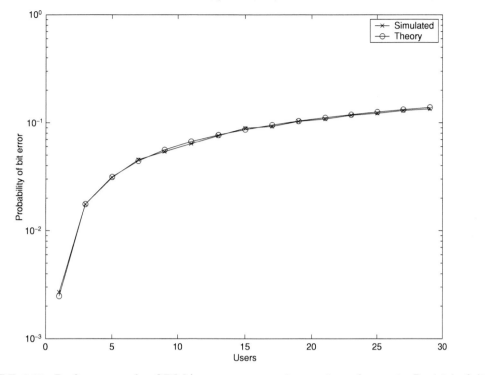

FIGURE 2.30: Performance of a CDMA system versus the number of users in Rayleigh fading $(\overline{E_b/N_0} = 20\text{dB}$, flat fading).

is non-coherent FSK. In an AWGN channel and in the absence of MAI, the link performance (in terms of symbol error rate) of FS/SS using non-coherent M-FSK modulation/reception can be shown to be

$$P_s = \sum_{k=1}^{M-1} \binom{M-1}{k} \frac{(-1)^{k+1}}{k+1} \exp\left(-\frac{k}{k+1}\frac{E_s}{N_0}\right) \qquad (2.84)$$

where M is the number of symbols and E_s/N_0 is the ratio of energy per symbol and noise power spectral density [22]. The bit error probability can be approximated as

$$P_e \approx \frac{1}{2}P_s \qquad (2.85)$$

and the energy per bit is equal to $E_b = E_s/\log_2 M$. This is simply the performance for non-coherent M-FSK in an AWGN channel with non-coherent reception. The performance with $M = 2, 4, 8$ is shown in Figure 2.31. We can see that, as is typical in orthogonal modulation schemes, the performance of FH/SS improves as the number of symbols increases. However, the performance improvement decreases rapidly with increasing M.

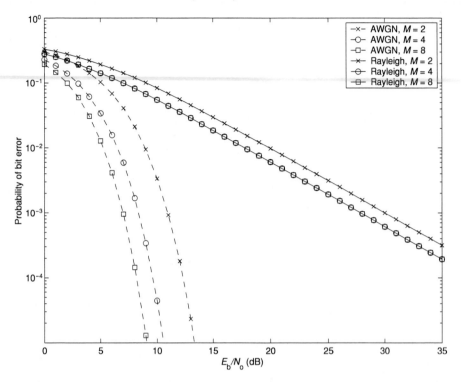

FIGURE 2.31: Performance of non-coherent FH/SS with M-FSK modulation in AWGN and flat Rayleigh fading (slow hopping).

In the presence of flat Rayleigh fading with slow hopping, each symbol undergoes fading and the performance in the absence of coding can be shown to be [1]

$$P_e = \frac{M}{2(M-1)} \sum_{k=1}^{M-1} \binom{M-1}{k} \frac{(-1)^{k+1}}{1 + k\left(1 + \left(\log_2 M\right) E_b/N_0\right)} \qquad (2.86)$$

where E_b/N_0 is the average energy per bit divided by the noise power spectral density where the expectation is over the Rayleigh fading. The performance for $M = 2, 4, 8$ is again plotted in Figure 2.31. We can see that the first Rayleigh fading, as in the DS/SS case, causes a dramatic performance degradation as compared to the AWGN channel. Additionally, we can see that increasing M does not improve performance as much as in the AWGN channel. Again, the performance of FH/SS is no different than of standard non-coherent M-FSK. However, spread spectrum has an advantage over narrowband modulation in that it can obtain diversity through bandwidth expansion. We have already seen this in the case of DS/SS. With FH/SS, the signal is hopped across a bandwidth much larger than the original signal bandwidth. Ideally, the resulting bandwidth is substantially larger than the coherence bandwidth of the channel (i.e., the range of frequencies over which the received signal is highly correlated). When this is accomplished, symbols transmitted during different hop periods will experience different fading conditions. The most straightforward means of obtaining diversity in such a situation is through fast hopping, i.e., hopping multiple times per symbol. If this is accomplished and the distance between hops (in terms of frequency) is greater than the coherence bandwidth of the channel, diversity is obtained during every symbol interval. As in DS/SS, the diversity improves performance since it provides multiple opportunities for a good channel. The level of diversity directly depends on both the number of hops per symbol and the coherence bandwidth of the channel. Specifically, the diversity level L will be approximately equal to

$$L \approx \min \left(\left\lfloor \frac{B_s}{B_c} \right\rfloor, \left\lfloor \frac{T_s}{T_c} \right\rfloor \right) \qquad (2.87)$$

where B_s is the spread (or system) bandwidth, B_c is the coherence bandwidth, T_s is the symbol duration, T_c is the hop period, and $\lfloor x \rfloor$ represents the floor function (the largest integer smaller than x). With L-fold diversity, the performance of non-coherent BFSK ($M = 2$) can be written as [1]

$$P_e = \left(\frac{L}{2L + E_b/N_0} \right)^L \sum_{k=0}^{L-1} \binom{L-1+k}{k} \left(1 - \frac{L}{2L + E_b/N_0} \right)^k \qquad (2.88)$$

The performance of $L = 1, 2, 4, 8$ is plotted in Figure 2.32. We can easily see that, as in the case of DS/SS, the diversity afforded by spread spectrum provides tremendous peformance enhancement in Rayleigh fading. Unfortunately, one difference in FH/SS is the difficulty of fast hopping. Hopping frequencies multiple times per symbol can be an implementation challenge. An alternative is to employ error correction coding and interleaving. This allows the temporal diversity afforded by frequency hopping to be incorporated into the coding decision statistics.

2.8   MULTIPLE ACCESS PERFORMANCE OF FREQUENCY-HOPPED CODE DIVISION MULTIPLE ACCESS

Although the emphasis of this chapter is DS-CDMA, we would like to examine the performance of FH-CDMA systems. In general, frequency hopping can provide many of the same performance benefits that DS-CDMA can provide, including diversity in fading channels and interference averaging [often termed interferer diversity in frequency-hopped multiple access (FHMA) contexts]. To explore the performance in relation to the system loading (i.e., the number of users accessing the system), we first make a few simplifying assumptions: (a) we will examine the uplink, (b) each signal hops once per symbol, (c) BFSK modulation is assumed with non-coherent reception, (d) frequency hop patterns are independent and identically

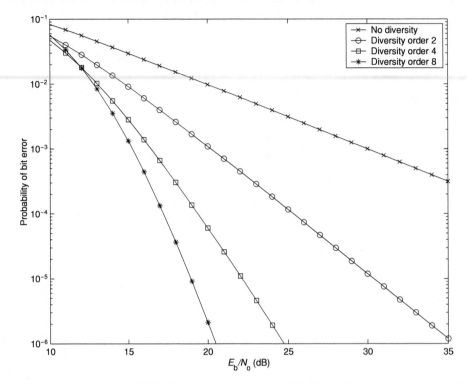

FIGURE 2.32: Performance of FH/SS with non-coherent BFSK modulation with various levels of diversity.

distributed and are assumed random and memoryless, and (e) perfect power control is achieved at the receiver.

The probability of error for FH-CDMA can be written as [35]

$$P_e = P_o (1 - P_h) + P_1 P_h \qquad (2.89)$$

where P_o is the probability of error when there are no collisions (or *hits*) between users (i.e., when two users avoid hopping to the same frequency at the same time), P_h is the probability of at least one hit, and P_1 is the probability of error when at least one hit has occurred. With BFSK and non-coherent reception, the probability of error in the absence of collisions is [22]

$$P_o = \frac{1}{2} \exp\left(-\frac{1}{2}\frac{E_b}{N_0}\right) \qquad (2.90)$$

Now, with independent random hopping codes, the probability of at least one hit can be written as

$$P_h = 1 - (1 - P_2)^{K-1} \qquad (2.91)$$

where P_2 is the probability of any two users hopping to the same frequency in the same symbol duration. For independent random hop codes, this probability is $P_2 = 1/N$ for synchronous hopping and $P_2 = 2/N$ for asynchronous hopping [1].

We will examine the problem by determining upper and lower bounds for the bit error probability. Clearly, an upper bound can be formed by assuming that the probability of error whenever a hit occurs is 50%, i.e., $P_1 = 1/2$ [35]:

$$P_b \leq P_o (1 - P_h) + \frac{1}{2} P_h \qquad (2.92)$$

A lower bound can be determined by examining the case when a collision between the desired user and exactly one other user occurs. Clearly, this probability is less than the probability of one or more collisions occurring, i.e.,

$$P_h < (K - 1) P_2 (1 - P_2)^{K-2} \qquad (2.93)$$

Further, the probability of error when a collision occurs between two users depends on the relative value of the bits. If the two values are the same (which occurs with probability 1/2), the probability of error is P_o. If the two values are different, the probability of error is 1/2. This leads to a lower bound

$$P_e \geq P_o (1 - P_h) + \frac{1}{2} \left(\frac{1}{2} + P_o \right) (K - 1) P_2 (1 - P_2)^{K-2} \qquad (2.94)$$

This lower bound tends to be accurate whenever K is small relative to N since it assumes that the probability of multiple hits is negligible when compared to hits from one signal. An example is plotted in Figure 2.33 for $N = 100$, $E_b/N_0 = 10$, and synchronous hopping. The figure contains plots for the upper and lower bounds as well as simulation results. We can see that the upper bound is overly pessimistic but the lower bound is fairly accurate. Again, the lower bound will be more loose for higher loading factors. Improved bounds were developed [36] and are particularly helpful in high loading conditions.

One advantage that FH-CDMA has over DS-CDMA is its resistance to the near-far problem. Specifically, examining (2.94) and the equations leading up to it, we see that the performance is independent of the SIR because we have made the slightly pessimistic assumption that any collision results in a 50% BER. This will not always be the case, but it is certainly a good approximation as we will see next. Thus, regardless of the SIR, the impact of collisions between users on the performance is the same. This can be seen in Figure 2.34, which plots the simulated BER for FH-CDMA along with the upper and lower bounds with non-coherent BFSK, $N = 100$ frequencies, and $K = 25$ users. The performance is plotted versus the near-far ratio or the power of one interfering signal compared to the desired signal. As the single interferer grows very strong relative to the desired signal, performance is essentially unaffected.

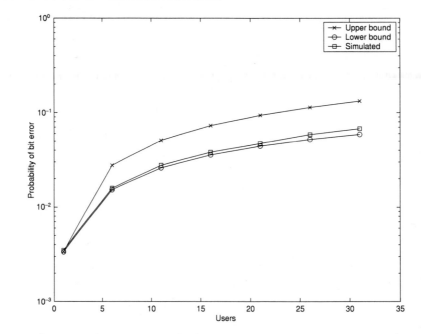

FIGURE 2.33: System performance for FHMA system with random hopping codes (100 frequencies, non-coherent BFSK).

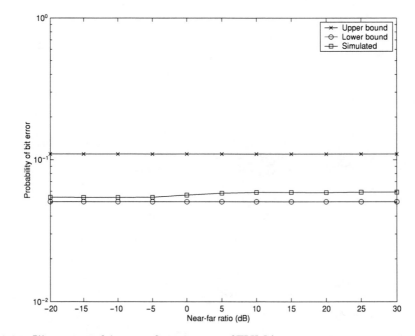

FIGURE 2.34: Illustration of the near-far resistance of FHMA.

This is in stark contrast to DS-CDMA, which is very sensitive to the near-far problem as seen in Figure 2.29.

2.9   SUMMARY

In this chapter, we have described the two basic spread spectrum techniques used in CDMA systems : frequency hopping and direct sequence. We also described the performance of the techniques in a single-user environment as well as the impact of multiple simultaneous transmissions also known as MAI. We showed that in fading environments, spread spectrum signals provide a substantial performance advantage over narrowband signals due to the frequency diversity that can be harnessed. However, in a multiple user environment, MAI can severely degrade performance and is the limiting performance factor, particularly in the presence of large received power disparities. In the following chapter, we will examine a cellular environment that exploits the properties of spread spectrum signals to improve the overall system capacity despite the limitations due to MAI.

CHAPTER 3

Cellular Code Division Multiple Access

In the previous chapter, we examined direct sequence and frequency-hopped forms of spread spectrum in both single-user and multiuser environments. One of the most prominent uses of CDMA is in cellular applications. In fact, cellular applications were the spring board from which spread spectrum made the leap from a military technology to a commercial technology. Thus, in this chapter, we specifically discuss cellular CDMA systems, focusing on direct sequence. We will first describe four basic principles of CDMA cellular systems that distinguish them from cellular systems based on other multiple access techniques. We will then examine the capacity of cellular CDMA systems, which relies heavily on these basic principles. Finally, we will discuss radio resource management, the primary system-level function of cellular CDMA systems.

3.1 PRINCIPLES OF CELLULAR CODE DIVISION MULTIPLE ACCESS

In this section, we will discuss four key principles of CDMA systems, particularly cellular systems: interference averaging, statistical multiplexing, universal frequency reuse, and soft hand-off. Each of these characteristics is both a fundamental advantage of CDMA systems and derives from the channels' sharing of a single frequency band and time slot.

3.1.1 Interference Averaging

The first principle that we will discuss is interference averaging. Recall that in-band interference is inevitable in all wireless systems. In TDMA/FDMA systems, we attempt to minimize this interference by separating co-channel signals by a sufficient distance. Typically, a signal may experience interference from two to five co-channel signals. As discussed in Chapter 2, a key aspect of spread spectrum is spreading gain. The despreading process mitigates the interference that any one signal causes to another signal. As a result, in CDMA, we increase the number of interfering signals in exchange for reducing the impact that any one signal has. (In TDMA or FDMA, each channel will see one to seven interfering signals. In CDMA, each channel will see

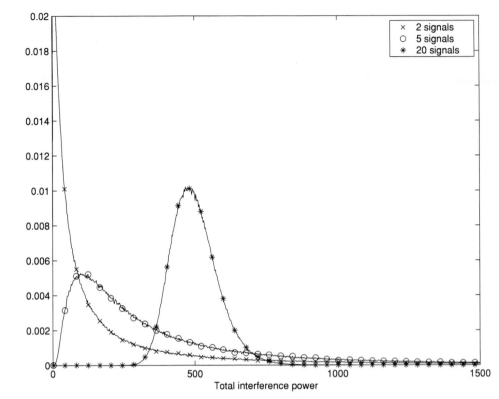

FIGURE 3.1: PDF of interference with varying number of log-normally distributed interferers.

dozens of interfering signals.) To understand this more clearly, consider the interference power caused by two log-normally distributed interferers each with a mean value of 250 (the units are irrelevant for the nature of the discussion) so that the total interference has a mean value of 500 and a standard deviation of approximately 3342. The probability density function is shown in Figure 3.1. If the interferences were due to five interferers each with a mean received power of 100, the mean of the total interference does not change, but the standard deviation is reduced (1306) as can be seen in Figure 3.1. Now, if we increase the number of signals to twenty and reduce the average power of each to 25, the average total interference remains the same, but the standard deviation is reduced dramatically ($\sim$83) as seen in Figure 3.1. Two effects occur: the distribution tends toward a Gaussian distribution due to the Central Limit Theorem, and, more importantly, the variance is reduced significantly due to the law of large numbers.

So exactly how does this benefit us? The performance of a wireless system is directly dependent on the signal-to-interference ratio (SINR), which is a random variable due to the random interference (ignoring for the moment the variation of the desired signal due to fading). The performance of a wireless link can be viewed in terms of either its average BER value or the probability that the SINR drops below a desired threshold (termed *outage probability*). In either case, the performance is dominated by the tails of the SINR distribution. Because the tails are shortened by reducing the variance, the required average value to obtain a target outage probability is reduced. We will solidify this idea through the following example.

Example 3.1.   Consider a wireless system where the received signal power at the edge of the coverage area due to power control is log-normally distributed with a mean value of -110dBm and a standard deviation of 1dB. The interference due to a single dominant co-channel interferer varies depending on the position of the interferer. It can also be modeled as a log-normal random variable with a mean value of -130dBm and a standard deviation of 6dB. What SIR value is exceeded 99% of the time? If the interference were instead composed of 50 signals each with 1/50th of the power of the original interference, what SIR is exceeded 99% of the time?

Solution: The SIR is the ratio of the desired signal power S to the interference power I. Since both are log-normal random variables, it is easy to show that the ratio S/I is also a log-normal random variable with parameters $\mu = \mu_S - \mu_I$ and $\sigma = \sqrt{\sigma_S^2 + \sigma_I^2}$ (where $\mu_S = E\{\ln(S)\}$, $\mu_I = E\{\ln(I)\}$, $\sigma_S^2 = \text{var}\{\ln(S)\}$, and $\sigma_I^2 = \text{var}\{\ln(I)\}$) and the probability density function can be written as

$$f_{SIR}(x) = \frac{1}{x\sigma\sqrt{2\pi}}\exp\left(-\frac{(\ln(x) - \mu)^2}{2\sigma^2}\right) \qquad (3.1)$$

In base 10, the mean of the SIR is -110dBm $+ 130$dBm $= 20$dB and the standard deviation is $\sqrt{1 + 36} = 6.08$dB. Converting to base e,

$$\mu = \frac{20}{10}\ln(10)$$
$$= 4.61 \qquad (3.2)$$
$$\sigma = \frac{6.08}{10}\ln(10)$$
$$= 1.40 \qquad (3.3)$$

Now we wish to find X such that

$$\Pr\{SIR \geq X\} = 0.99 \qquad (3.4)$$

Using the Q-function defined in Chapter 2,

$$0.01 = Q(-X)$$
$$= Q\left(\frac{\ln(x) - \mu}{\sigma}\right) \qquad (3.5)$$

$$-\frac{\ln(x) - \mu}{\sigma} = 2.33 \qquad (3.6)$$
$$x = \exp(-2.33\sigma + \mu)$$
$$= 3.83$$
$$= 5.8\text{dB} \qquad (3.7)$$

Thus, the system maintains a 5.8-dB SIR or better 99% of the time. Now, let us examine the case when the interference is made up of fifty independent signals. The exact distribution of the sum of log-normal random variables is unknown. However, the sum can be approximated as a log-normal distribution [37, 38]. Specifically, consider the random variable

$$\Gamma = \frac{1}{K}\sum_{i=1}^{K} \lambda_i \qquad (3.8)$$

where λ_i are independent identically distributed log-normal random variables with parameters μ_λ and σ_λ. Γ can be approximated by a log-normal random variable with [38]

$$\mu_\Gamma = \mu_\lambda + \frac{\sigma_\lambda^2}{2} + \ln\left(1 \left/ \left(\sqrt{1 + \frac{\exp(\sigma_\lambda^2) - 1}{K}}\right)\right.\right) \qquad (3.9)$$

$$\sigma_\Gamma^2 = \ln\left(1 + \frac{\exp(\sigma_\lambda^2) - 1}{K}\right) \qquad (3.10)$$

Substituting values for μ_λ and σ_λ, we have $\mu_\Gamma = -29.3$ and $\sigma_\Gamma = 0.42$. Converting back to log base 10 and using the formula for the ratio of two log-normal random variables, the SIR is log-normal with a mean value of 16dB and standard deviation of 1.75dB, which in base e is $\mu = 3.7$ and $\sigma = 0.42$. Substituting into (3.7), we have

$$x = \exp(-2.33\sigma + \mu)$$
$$= 15.9$$
$$= 12.0\text{dB} \qquad (3.11)$$

Thus, the system with interference averaging has a 6.2dB higher SIR than the system without interference averaging. Figure 3.2 shows the CDF using the log-normal approximation and the simulated cumulative histogram. We can see that the log-normal approximation is very accurate.

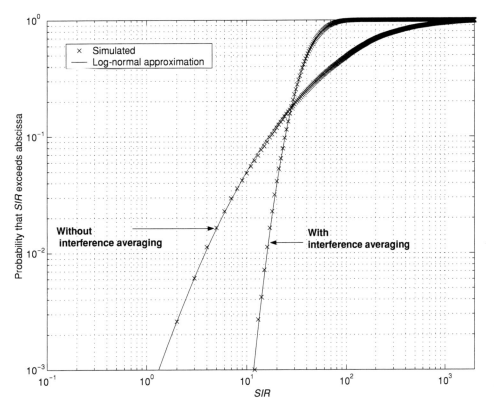

FIGURE 3.2: Simulated and theoretical CDFs for SIR with a single interferer and with interference averaging.

3.1.2 Frequency Reuse

As discussed in Chapter 1, frequency reuse is an important concept in wireless systems, particularly cellular systems. Propagation losses allow frequency bands to be reused in geographically separated locations. This increases the overall capacity of wireless systems. However, in a given area, frequency reuse means that only a fraction of the total number of channels are available with the fraction being inversely related to the frequency reuse pattern. For example, with a frequency reuse pattern of $Q = 7$, as illustrated in Figure 1.4, the total number of channels available in a given area is $C = N_{tot}/7$ where N_{tot} is the total number of channels available. In an FDMA system, the total number of channels is proportional to the total available bandwidth divided by the desired data rate per user, $N_{tot} \propto B/R_b$, or the total number of dimensions. In a TDMA system, the number of time slots available is also equal to the total number of dimensions available: $N_{tot} \propto B/R_b$. Thus, the total number of channels available is the same as in FDMA.

One of the primary advantages of CDMA systems is universal frequency reuse. That is, since the waveform is designed to tolerate interference, we can reuse all frequencies in each cell. On first glance, one might expect an automatic improvement in the capacity of CDMA systems by a factor of Q, the reuse factor, as compared to TDMA or FDMA systems. However, this is not exactly the case. For CDMA to take full advantage of frequency reuse, systems would have to use the entire available number of dimensions in each cell. In CDMA systems, the dimensionality of the signal is increased by N, which is the spreading factor or bandwidth expansion factor. To fully realize a capacity gain of Q, the system would need to support N channels per NR_b Hz of bandwidth. This is impossible due to both in-cell and out-of-cell interference.

In Chapter 2, we found that the SINR of a CDMA system can be approximated using a Gaussian assumption on the interference. Specifically, the SINR on the uplink of a CDMA system can be written as

$$SINR = \frac{N * S}{\sum_{i=2}^{K} I_i + \sigma_n^2} \tag{3.12}$$

where K is the number of in-cell interferers, S is the desired signal power, I_i is the ith interferer's power, N is the processing gain, and σ_n^2 is the thermal noise power. With perfect power control and ignoring thermal noise, we can write

$$SIR = \frac{N}{K - 1} \tag{3.13}$$

This expression ignores out-of-cell interference. It can be shown that the out-of-cell interference can also be modeled as a Gaussian random variable with a variance that is a factor of the in-cell interference power. Thus, including out-of-cell interference, we have

$$SIR = \frac{N}{(K - 1)(1 + f)} \tag{3.14}$$

where $(1 + f)$ accounts for the out-of-cell interference.

If we require a specific SIR to achieve the target performance, we can solve for the total number of channels possible:

$$K = \frac{N}{SIR(1 + f)} + 1$$
$$\approx \frac{N}{SIR(1 + f)} \tag{3.15}$$

Since the bandwidth expansion factor N is equal to the bandwidth divided by the data rate, we have

$$K_{cdma} \approx \frac{R_b}{B} \frac{1}{SIR(1 + f)} \tag{3.16}$$

Comparing this to TDMA or FDMA,

$$K_{tdma} \approx \frac{R_b}{B}\frac{1}{Q} \qquad (3.17)$$

and we can see that the effective reuse factor of CDMA is $Q_{eff} = SIR(1+f)$. Thus, while CDMA does provide universal frequency reuse, it fails to translate directly into a factor Q improvement over a system with a reuse factor of Q. Rather, it provides an improvement of $Q/(SIR(1+f))$, which may be less substantial. However, universal frequency reuse does provide other benefits. A primary benefit is the possibility of soft hand-off, which is discussed next.

3.1.3   Soft Hand-Off

Due to the limited range of a given base station (or access point in a WLAN context), if a wireless transmitter/receiver moves, it may exceed the coverage. When this happens, a different base station must act as the access point to the network. The process of changing base stations during a call (or session) is termed *hand-off* and is a fundamental aspect of mobile communication systems. In traditional TDMA or FDMA systems, a hard hand-off must take place where the mobile ceases communication with its current base station before beginning communication with a new base station. This "break-before-make" mode of hand-off is necessary if only a single RF channel can be monitored at a time (or if we want to avoid using two channels per mobile) and results in the possibility of a dropped call during hand-off. A third fundamental characteristic of CDMA systems is the use of *soft hand-off*.

Universal frequency reuse facilitates the use of soft hand-off. In soft hand-off, the mobile makes a connection with the new base station *before* breaking a connection with the old base station. This "make-before-break" approach substantially reduces the probability of a call drop during hand-off. Because of universal frequency reuse, the same RF band can be monitored for both base stations with the channels being separated through the despreading process in baseband processing.

In addition to reducing the call dropping probability, the major advantage of soft hand-off is that it provides *macro-diversity*. Recall from the previous chapter that spread spectrum systems provide inherent diversity versus the deleterious effects of multipath fading. However, spread spectrum systems (like all wireless systems) are also subject to log-normal shadowing. As mentioned previously, log-normal shadowing is the variation in received signal strength resulting from macroscopic effects such as buildings and geographic features. As a result, a Rake receiver, while providing robustness against multipath fading, is ineffective against log-normal macroscopic fading. However, the use of soft hand-off allows for diversity against log-normal shadowing. Since the mobile is communicating with two base stations simultaneously, if the signal provided by one base station enters into a shadowed region, a strong probability exists that the other signal will not be shadowed.

Example 3.2.   Consider a situation where a mobile unit is on the edge of the cell boundary and is thus equidistant to two base stations. The SNR from one base station follows a log-normal distribution with $\mu = 20$dB and $\sigma = 12$dB. What is the probability that the SNR is less than 6dB? If the system uses soft hand-off with maximal ratio combining between the two base stations with half the power from each, what is the probability that the SNR is less than 6dB?

Solution: The SNR without soft hand-off is a log-normal random variable with mean 20dB and standard deviation 12dB. Now, instead of converting to base e as we did before, let us perform all calculations in base 10. The probability that the SNR exceeds 6dB is

$$\Pr\{SIR\} \geq 10^{6/10} = 1 - Q\left(\frac{\mu - 6}{\sigma}\right)$$
$$= 1 - Q(1.15)$$
$$= 0.88 \qquad (3.18)$$

Thus, the probability of exceeding the threshold is only 88%. Examining the soft hand-off case, unfortunately the distribution for the sum of two log-normal random variables is unavailable in closed form. However, it can be shown that for two log-normal random variables with parameters μ_1, σ_1 and μ_2, σ_2, the sum has an average value of [39]

$$\overline{\gamma}_{MRC} = \exp\left(\mu_1 + \frac{\sigma_1^2}{2}\right) + \exp\left(\mu_2 + \frac{\sigma_2^2}{2}\right)$$
$$= 2\overline{\gamma} \qquad (3.19)$$

where $\overline{\gamma}$ is the average SNR per soft hand-off leg (it is assumed to be the same for both legs in this case). If the power from each base station is reduced by half, the average SNR is unchanged by soft hand-off. Further, it can be shown that the variance of the SNR of maximal ratio combining is also additive [39]:

$$\sigma_{MRC}^2 = \exp\left(2\mu_1 + \sigma_1^2\right)\left(\exp\left(\sigma_1^2\right) - 1\right) +$$
$$\exp\left(2\mu_2 + \sigma_2^2\right)\left(\exp\left(\sigma_2^2\right) - 1\right)$$
$$= 2\sigma_{\gamma}^2 \qquad (3.20)$$

where σ_{γ}^2 is the variance of the SNR per hand-off leg. Additionally, one can easily show that if γ is a log-normal random variable, then

$$E\left\{\frac{\gamma}{2}\right\} = \frac{\overline{\gamma}}{2} \qquad (3.21)$$

$$\text{var}\left\{\frac{\gamma}{2}\right\} = \frac{\sigma_{\gamma}^2}{4} \qquad (3.22)$$

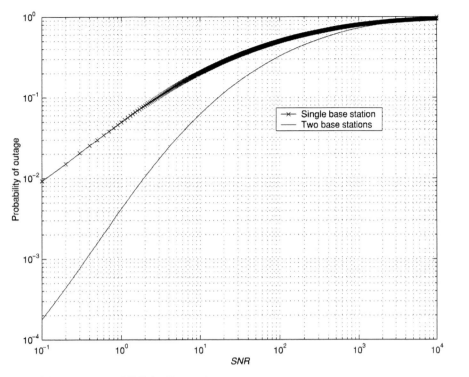

FIGURE 3.3: The empirical CDF for Example 3.2.

Thus, reducing the power on each hand-off leg by a factor of two will maintain an average SNR that is equal to the hard hand-off case but reduces the variance of the SNR by a factor of two relative to the hard hand-off case. To obtain the exact benefit, simulations can be run to determine the empirical CDF of the SNR. The result is shown in Figure 3.3, which plots the probability that the received SNR exceeds a given threshold obtained from simulation. We can see that in the hard hand-off case the probability that the SNR is less than 6dB (linear value of four) is approximately 12%, which matches our previous calculation. Examining the second curve, we find that for the soft hand-off case, the probability of not exceeding 6dB is roughly 2%. Thus, soft hand-off improves the coverage of CDMA systems.

The previous example demonstrated the benefits of soft hand-off from a macro-diversity perspective. By increasing the probability that the received SNR exceeds a required threshold, the system improves the coverage for a given amount of transmit power. In typical CDMA systems, a mobile can, in many situations, communicate with three base stations simultaneously, resulting in three-way soft hand-off. This provides further improvement, although, as with all diversity situations, the greatest benefit is achieved from the first diversity source [40].

3.1.4   Statistical Multiplexing

The final fundamental CDMA concept that we will discuss in this section is termed *statistical multiplexing*, which is the sharing of a single channel by multiple information streams. For example, ten 3-kbps information streams can be time-multiplexed onto a single 30-kbps channel. In time-multiplexing, the overall channel capacity is divided into time slots that can be used to serve each of the information streams, not unlike TDMA.

In voice-centric networks, despite the use of circuit-switched (i.e., dedicated) access to channel resources, any given transmitter sometimes has no data to send, and this is the motivation for statistical multiplexing. In a typical conversation, each user speaks approximately three-eighths of the time [41]. This fact is typically termed *voice activity* and can be exploited to increase the overall capacity of a CDMA system. Essentially, voice activity means that a given channel is unused roughly 60% of the time. The most straightforward means of exploiting this fact is to reassign a channel whenever it is unused. However, this is extremely impractical since voice inactivity periods are only insignificantly longer than the time required to reassign the channel.

Fortunately, in CDMA systems, reassigning the channels to take advantage of voice activity is unnecessary. Instead, we can exploit voice activity at the transmitter simply by suppressing transmission (or reducing the power substantially) whenever a speaker is inactive for a particular time frame (20ms in current systems). By suppressing transmission during inactive periods, less interference is generated and a larger pool of transmitters can be supported. This can be seen in Figure 3.4 where the interference statistics for a specific signal (after despreading) are plotted

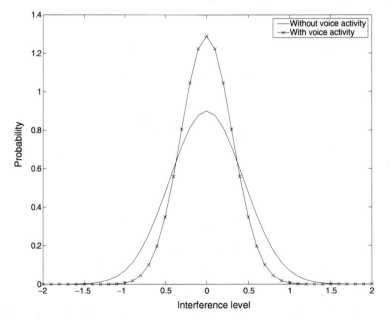

FIGURE 3.4: Example of interference statistics with and without voice activity ($K = 25$, $N = 64$, $\upsilon = 0.5$).

with and without voice activity. Specifically, we plot the second term on the right-hand side of equation (2.64) which is the interference portion of the decision statistic. The interference calculations assume random spreading codes, phases, delays, $K = 25$ users, a spreading factor of $N = 64$, and a voice activity factor of $\upsilon = 0.5$. It is clear that a voice activity of 50% reduces the variation of the interference by roughly a factor of two. Remember from Chapter 2 that the performance is inversely related to the interference variance; if we increase the number of users by a factor of two, we will maintain the original performance without voice activity. This is one of the primary advantages of CDMA. Since system resources are consumed primarily by interference, not by explicit channel usage, voice activity directly translates into larger capacity by reducing the amount of interference generated.

3.2   CODE DIVISION MULTIPLE ACCESS SYSTEM OVERVIEW

Now that the key concepts of CDMA have been introduced, a typical cellular CDMA system can be described. Here we are specifically interested in DS-CDMA systems, although many of the concepts and general trends are also applicable to FH-CDMA systems. First, we must distinguish between two distinct problems in CDMA systems: the uplink and the downlink.

The uplink is depicted in Figure 3.5 and is the link from the mobile station to the central base station. Synchronism between arriving signals is difficult to maintain, and thus signals are assumed to arrive asynchronously. The interference seen at the base station in the demodulation of any given signal due to in-cell and out-of-cell users is the sum of a large number of low-powered signals. Since transmitters are scattered throughout the cell, the path loss seen by the different users varies wildly. To ensure that the received signals from nearby users do not overwhelm the received signals from distant users (recall the near-far problem discussed in Chapter 2), tight power control is necessary (power control will be discussed in Section 3.4.1). Since the interference comprises a very large number of relatively low-powered

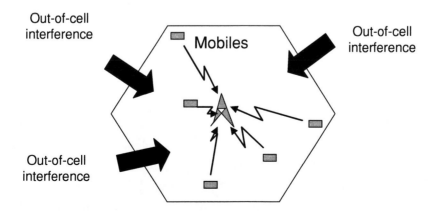

FIGURE 3.5: Illustration of the uplink of a cellular CDMA system.

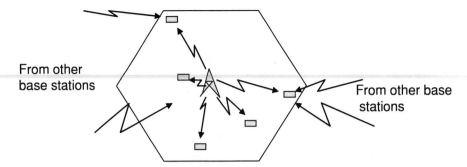

FIGURE 3.6: Illustration of the downlink in a cellular CDMA system.

interferers, it is well modeled by a Gaussian random process and the system benefits tremendously from interference averaging. The essential link design problem is one of interference management.

The downlink, on the other hand, is the link from the centralized base station to the various mobile stations in the cell as shown in Figure 3.6. Transmission of all signals occurs simultaneously, guaranteeing synchronous reception of all in-cell user signals at the mobile unit. In a flat fading channel, in-cell interference can be completely avoided through the use of orthogonal spreading codes. However, resolvable multipath is inevitable in most cellular CDMA systems, resulting in some amount of interference between in-cell channels on the downlink. Nevertheless, the reduction in in-cell interference mitigates the need for power control. In further contrast to the uplink, out-of-cell interference on the downlink occurs due to a small number of high-powered signals emanating from neighboring base stations. As a result, capacity is degraded because these is much less interference averaging. Another difference between the two links concerns soft hand-off. In the uplink, soft hand-off is accomplished simply by having multiple base stations listening to the transmitted signal from the mobile and relaying the decoded signal to the mobile switching center. However, on the downlink, orthogonal codes are used for transmitting. As a result, a limited number of channels are available, and soft hand-off consumes additional channels. The system design problem for the downlink, as opposed to the uplink, is a combined power and code management problem. Since there is a fixed amount of transmit power and a fixed number of orthogonal codes, both can limit capacity.

The differences between the two links results in different design strategies. CDMA system designs (e.g., IS-95) take advantage of the broadcast nature of the downlink by transmitting a strong common pilot for acquisition and coherent demodulation. Additionally, because synchronism can be maintained betweeen user signals, the downlink employs orthogonal spreading codes. Initial deployments of CDMA used slow power control since it was less necessary on the downlink. The uplink, in contrast, initially used non-coherent demodulation due to the lack of a pilot, employed long pseudo-random spreading codes, and required tight power control to avoid the near-far problem.

3.3   CAPACITY

In this section, we will discuss the capacity of CDMA systems with an eye toward providing a first level approximation of the capacity and the major factors that influence capacity. First, let us make an approximation based on average interference levels and the average required SINR. We will then examine the second-order statistics of the interference to determine the capacity based on the outage probability. Additionally, the second-order analysis will differentiate between the uplink and the downlink.

3.3.1   Comparison of Multiple Access Capacity

To see the advantage of CDMA over other multiple access techniques, let us first pursue a first-order comparison of the capacity of FDMA, TDMA, and CDMA based on average SINR performance. Let us assume that a system has a total bandwidth of B_THz and requires a bit rate of R_b per user. In an FDMA system, the bandwidth required per user depends directly on the modulation and pulse-shaping scheme used. For the sake of simplicity, let us assume a linear modulation scheme with one bit per symbol and optimal pulse shaping. In this case, the bandwidth required per user is $B_u = R_b$ and the system capacity (ignoring guard bands) is

$$K_{fdma} = \frac{B_T}{B_u} = \frac{B_T}{R_b} \qquad (3.23)$$

Now in a TDMA system, the overall transmission rate that can be supported R_b^S is equal to the total bandwidth (again making the same assumptions as above): $R_b^S = B_T$. The system capacity is the overall system data rate divided by the data rate required per user

$$K_{tdma} = \frac{R_b^S}{R_b} = \frac{B_T}{R_b} \qquad (3.24)$$

which is the same as FDMA. This is intuitive since it does not matter whether we slice the resources into frequency or time slices unless one is more easily implemented than the other. In a CDMA system, if time synchronism between users can be maintained, signals can be made orthogonal through the use of Walsh codes and the capacity is the same as in FDMA and TDMA.

The more interesting case is when synchronism cannot be maintained, as is typical on the uplink of a CDMA system. In this case, the signals are non-orthogonal and the system is interference limited. As was demonstrated in Chapter 2, performance is directly related to SINR. Specifically, the performance is directly related to the ratio of the energy per bit E_b and the interference power spectral density I_0. This ratio can be written as

$$\frac{E_b}{I_0} = \frac{PT_b}{\left(\sum_k P_k\right)/B_T}$$

$$= \frac{B_T/R_b}{(K-1)} \tag{3.25}$$

where we have made the assumption of perfect power control $P_k = P, \ \forall k$. Thus, the capacity of the CDMA system can be written as

$$K_{cdma} = \frac{B_T/R_b}{E_b/I_0} + 1$$

$$\approx \frac{B_T/R_b}{E_b/I_0} \tag{3.26}$$

Comparing this equation to the TDMA and FDMA analyses, we find that for a required $E_b/I_o > 1$, the capacity of CDMA is lower than that of TDMA or FDMA because the signals are non-orthogonal. However, we have yet to consider the effects of frequency reuse, voice activity, or sectorization. As discussed in Chapter 2 and earlier in this chapter, wireless systems typically employ frequency reuse to increase the capacity over a geographical area. However, this reduces the number of channels available in a given area. Assuming a frequency reuse factor of Q, the capacity of TDMA/FDMA is

$$K_{tdma/fdma} = \frac{1}{Q}\frac{B_T}{R_b} \tag{3.27}$$

Due to universal frequency reuse, the number of channels remains the same in CDMA, but the overall interference is increased, which results in an effective reuse factor as we saw in Section 3.1.2. Including out-of-cell interference, the capacity can be written as

$$K_{cdma} \approx \frac{B_T/R_b}{(1+f)\,E_b/I_0} \tag{3.28}$$

where f is the amount of out-of-cell interference as a fraction of the in-cell interference. The next factor that must be considered is the voice activity factor v. As was discussed in Section 3.1.4, voice activity reduces the interference seen by each signal due to statistical multiplexing. On average, the interference power is directly reduced by the voice activity factor v. This factor is unexploited in TDMA or FDMA, but the capacity of CDMA is directly increased by $1/v$.

The final factor that should be considered is the impact of sectorization, which is well understood to reduce the frequency reuse factor required to achieve a given SIR [7]. In a CDMA system, antenna sectorization reduces the interference by the antenna gain factor G. Including each of these effects results in system capacity of

$$K_{cdma} \approx \frac{G\,(B_T/R_b)}{\nu\,(1+f)\,E_b/I_0} \qquad (3.29)$$

With three-sector antennas, the standard TDMA/FDMA sectorization factor is $Q = 7$, resulting in a capacity of $K_{tdma/fdma} = B_T/(7R_b)$ per cell or $B_T/(21R_b)$ per sector. A typical E_b/I_0 requirement for CDMA is 6dB. Using a three-sector antenna gain of 4dB (including a 1-dB scalloping loss), an interference factor of $f = 0.6$, and voice activity factor of $\nu = 3/8$, the approximate capacity of CDMA per cell is $K_{cdma} \approx B_T/R_b$, which is approximately an order of magnitude of capacity improvement.

3.3.2   Second-Order Analysis

The previous discussion of the capacity of CDMA systems is slightly misleading. The analysis provides the average capacity assuming that all interference variables assume their average values. However, as we have discussed previously, due to log-normal shadowing, voice activity, and the random location of mobiles in their respective cells, the interference is a random variable. What we would like to calculate is the probability of outage, i.e., the probability that the SINR falls below a required value. Note that this approach, while intuitive for the uplink, is not particularly useful for the downlink. On the uplink, capacity depends on the interference observed, but on the downlink, capacity depends on the power expended per user. Thus, we will take a slightly different (though closely related) approach for the downlink. Both analyses closely follow the approach given in the seminal paper by Gilhousen, *et al.* [42].

Uplink Capacity

To determine uplink capacity, let us return to the expression for the SINR for user 1 assuming perfect power control:

$$SINR = \frac{P_1}{\sum_{k=2}^{K} P_k + I + \widetilde{N}}$$

$$= \frac{1}{(K-1) + I/P + \widetilde{N}/P} \qquad (3.30)$$

where there are K in-cell (or K_s in-sector) interferers, I is the total out-of-cell interference, and $\widetilde{N}$ is thermal noise [42]. Including the data rate and the bandwidth, we can write

$$\frac{E_b}{I_0} = \frac{1/R_b}{\left[(K-1) + I/P + \widetilde{N}/P\right]/B_T}$$

$$= \frac{B_T/R_b}{(K-1) + I/P + \widetilde{N}/P} \qquad (3.31)$$

where I_0 includes thermal noise power spectral density. Including the effect of voice activity, we have

$$\frac{E_b}{I_0} = \frac{B_T/R_b}{\sum_{k=2}^{K} \psi_k + I/P + \tilde{N}/P} \qquad (3.32)$$

where ψ_k is a binary random variable in the set $\{0, 1\}$, which represents the voice activity of the kth user.

Assuming perfect power control, $\tilde{N}/P$ is a constant, but $\sum_{k=2}^{K} \psi_k$ is a binomial random variable and I/P is a random variable representing out-of-cell interference. The out-of-cell interference is the sum of a large number of log-normal random variables that is well modeled as a Gaussian random variable. The probability of outage is simply the probability that instantaneous E_b/I_0 falls below $(E_b/I_0)_{req}$ needed for a desired performance:

$$P_{out} = \Pr \left\{ \frac{B_T/R_b}{\sum_{k=2}^{K} \psi_k + I/P + \tilde{N}/P} < \left(\frac{E_b}{I_0} \right)_{req} \right\}$$

$$= \Pr \left\{ \sum_{k=2}^{K} \psi_k + I/P > \frac{B_T/R_b}{(E_b/I_0)_{req}} - \frac{\tilde{N}}{P} \right\} \qquad (3.33)$$

Thus, we require the statistics of $\sum_{k=2}^{K} \psi_k + I/P$. To determine the statistics of I/P, we assume a log-distance path loss model with log-normal shadowing. That is, the received signal power from a mobile at its cell site d_m meters away is proportional to $10^{\xi_m/10} d_m^{-\kappa}$ where ξ_m is a log-normal random variable and κ is the path loss exponent. Consider a mobile that is d_m meters from its serving base station and d_o meters from the base station of interest. Assuming independent shadowing terms to the two base stations (ξ_m, ξ_o), the normalized interference caused to the base station of interest is

$$\frac{I(d_o, d_m)}{P} = \frac{10^{\xi_o/10} d_o^{-\kappa}}{10^{\xi_m/10} d_m^{-\kappa}}$$

$$= \left(\frac{d_m}{d_o} \right)^{\kappa} 10^{(\xi_o - \xi_m)/10} \qquad (3.34)$$

which must be less than unity since each mobile is served by the base station with the strongest signal (i.e., no interfering base station can be stronger than the serving base station). Consider a single sector of a three-sector cellular system as shown in Figure 3.7. To find the total interference caused in the sector of interest due to out-of-cell mobiles, we assume a uniform density of users $\rho = 2K/(3\sqrt{3}) = 2K_s/\sqrt{3}$ in the hexagonal area and integrate over the area indicated in Figure 3.7.

The total interference experienced in the sector of interest is then

$$\frac{I}{P} = \int \int \psi \left(\frac{d_m}{d_o} \right)^{\kappa} 10^{(\xi_o - \xi_m)/10} \chi \left(\frac{d_m}{d_o}, \xi_o - \xi_m \right) \rho \, dA \qquad (3.35)$$

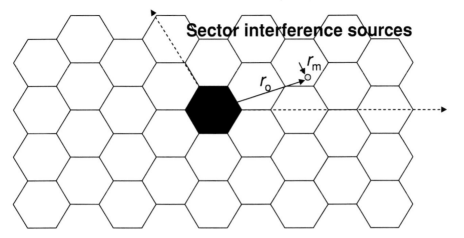
Sector interference sources

FIGURE 3.7: Illustration of out-of-cell interference calculation.

where

$$\chi\left(\frac{d_m}{d_o}, \xi_o - \xi_m\right) = \begin{cases} 1 & \left(\frac{d_m}{d_o}\right)^\kappa 10^{(\xi_o-\xi_m)/10} \leq 1 \\ 0 & \text{otherwise} \end{cases} \quad (3.36)$$

guarantees that only out-of-cell mobiles are included in the interference calculation and ψ is a voice activity random variable that is 1 with probability ν and 0 with probability $1 - \nu$. We wish to model I/P as a Gaussian random variable and thus require the mean and variance to completely describe it. The mean is found as

$$E\left\{\frac{I}{P}\right\} = E\left\{\int\int \psi\left(\frac{d_m}{d_o}\right)^\kappa 10^{(\xi_o-\xi_m)/10}\chi\left(\frac{d_m}{d_o}, \xi_o - \xi_m\right)\rho\, dA\right\}$$

$$= \int\int E\{\psi\}\left(\frac{d_m}{d_o}\right)^\kappa E\left\{10^{(\xi_o-\xi_m)/10}\chi\left(\frac{d_m}{d_o}, \xi_o - \xi_m\right)\right\}\rho\, dA$$

$$= \int\int \nu\left(\frac{d_m}{d_o}\right)^\kappa\left[\int_{-\infty}^{10*\kappa\log(d_m/d_o)} e^{x\ln(10)/10}\frac{e^{-x^2/4\sigma^2}}{\sqrt{4\pi\sigma^2}}dx\right]\rho\, dA$$

$$= \int\int \nu\left(\frac{d_m}{d_o}\right)^\kappa e^{[\sigma\,\ln(10)/10]^2}$$

$$\cdot\left[1 - Q\left(\frac{10*\kappa\log(d_m/d_o)}{\sqrt{2\sigma^2}} - \sqrt{2\sigma^2}\frac{\ln(10)}{10}\right)\right]\rho\, dA \quad (3.37)$$

where σ is the parameter of the log-normal random variable. By inserting values, we can determine the expected value through numerical integration. For example, for $\kappa = 4$, $\nu = 3/8$, and $\sigma = 8$dB, we obtain [42]

$$E\left\{\frac{I}{P}\right\} = 0.247K_s \quad (3.38)$$

In a similar fashion, we can show that

$$\text{var}\left\{\frac{I}{P}\right\} = \int\int \left(\frac{d_m}{d_o}\right)^{2\kappa} e^{[\sigma \ln(10)/10]^2}$$

$$\cdot\left[1 - Q\left(\frac{10 * \kappa \log(d_m/d_o)}{\sqrt{2\sigma^2}} - \sqrt{2\sigma^2}\frac{\ln(10)}{10}\right)\right] \qquad (3.39)$$

Referring back to (3.33), we wish to find the probability of outage for specific numbers of users per sector. Rewriting, we have

$$P_o = \sum_{k=0}^{K_s-1} \text{Pr}\left\{\frac{I}{P} > \frac{B_T/R_b}{(E_b/I_0)_{req}} - \frac{\tilde{N}}{P} - k \,\bigg|\, \sum_i \psi_i = k\right\}$$

$$\cdot\,\text{Pr}\left\{\sum_i \psi_i = k\right\}$$

$$= \sum_{k=0}^{K_s-1} \binom{K_s - 1}{k} v^k (1 - v)^{K_s-1-k} Q$$

$$\cdot\left(\frac{(B_T/R_b)/\left((E_b/I_0)_{req}\right) - (\tilde{N}/P) - k - 0.247K_s}{\sqrt{0.078K_s}}\right) \qquad (3.40)$$

The outage probability is plotted in Figure 3.8 for $B_T = 1.25\text{MHz}$, $R_b = 8\text{kbps}$ and $v = 3/8$. For a 1% outage probability, the system can support 36 users per sector or 108 users per cell. Comparing this with our previous simplistic analysis, which showed $K_{cdma} \approx B_T/R_b = 156$ users per cell, we find that the current estimate is significantly more conservative but is still approximately five times what we predicted for TDMA/FDMA schemes.

Downlink

The previous discussion examined the uplink, which, as discussed, is substantially different from the downlink. Obviously, uplink capacity is most useful when paired with similar downlink capacity. Thus, we wish to find the capacity of the downlink as well. Whereas uplink capacity is primarily concerned with interference power, the downlink is primarily concerned with transmit power.

Again following Gilhousen's development [42], assume that a mobile unit sees M base stations with relative powers $P_{T_1} > P_{T_2} > P_{T_3} > \cdots > P_{T_M} > 0$. Cell site selection is based on the base station with the strongest received power. The received E_b/I_0 at the mobile is lower bounded by

$$\frac{E_b}{I_0} \geq \frac{\beta f_i P_{T_1}/R_b}{\left[\left(\sum_{j=1}^{M} P_{T_j}\right) + \tilde{N}\right]/B_T} \qquad (3.41)$$

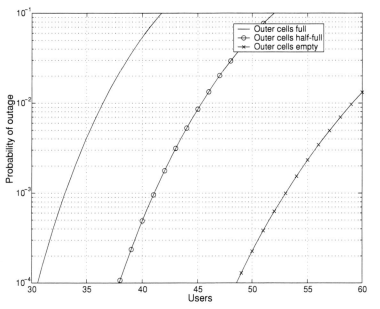

FIGURE 3.8: Uplink capacity in terms of outage probability with various loading levels.

where β is the fraction of the base station power devoted to the traffic [the common pilot signal used for acquisition and coherent demodulation is given $(1 - \beta)$ of the power] and f_i is the fraction of the traffic power devoted to the user of interest. Now supposing the required $(E_b/I_0)_{req}$ is given, the required transmit power fraction is upper bounded by

$$f_i \leq \frac{(E_b/I_0)_{req}}{\beta B_T/R_b} \left[1 + \left(\frac{\sum_{j=2}^{M} P_{T_j}}{P_{T_1}} \right)_i + \frac{\tilde{N}}{(P_{T_1})_i} \right] \qquad (3.42)$$

Since the total power devoted to the traffic cannot exceed what is available, we are constrained by

$$\sum_{i=1}^{K} f_i \leq 1 \qquad (3.43)$$

Defining the relative cell site powers as

$$\varphi_i \equiv \left(1 + \frac{\sum_{j=2}^{M} P_{T_j}}{P_{T_1}} \right)_i \qquad (3.44)$$

and summing over φ_i, we have the constraint

$$\sum_{i=1}^{K} \varphi_i \leq \frac{\beta B_T/R_b}{(E_b/I_0)_{req}} - \sum_{i=1}^{K} \frac{\tilde{N}}{P_{T_i}}. \qquad (3.45)$$

Defining the right-hand side of the inequality as δ, an outage occurs if the required relative powers exceed the limit given in (3.45). That is,

$$P_o = \Pr\left\{\sum_{i=1}^{K} \varphi_i > \delta\right\} \qquad (3.46)$$

Unfortunately, the distribution of φ_i does not lend itself to analysis. Following Gilhousen *et al.* [42], we simulated this value, and the histogram of $\varphi_i - 1$ is shown in Figure 3.9. From the histogram, we can compute the Chernoff bound on the outage probability as

$$\begin{aligned}
P_o &< \min_{\lambda>0} E\left\{\exp\left(\lambda \sum_{i=1}^{K} \varphi_i - \lambda\delta\right)\right\} \\
&= \min_{\lambda>0}\left[(1-v) + v\sum_{i=1}^{K} P_i \, \exp\left(\lambda\varphi_i\right)\right]^{K} \exp\left(-\lambda\delta\right)
\end{aligned} \qquad (3.47)$$

where P_k is the histogram value of φ in the kth bin.

For $R_b = 8\text{kbps}$, $E_b/I_0 = 5\text{dB}$, $B_T = 1.25\text{MHz}$, $\beta = 0.8$, and an SNR of -1dB, the resulting outage probability is plotted in Figure 3.10. For the same parameters, we can see that

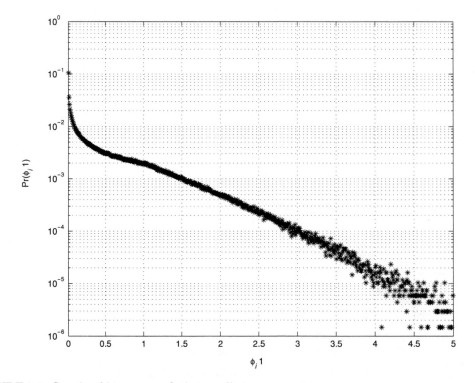

FIGURE 3.9: Simulated histogram of relative cell-site powers ϕ_i.

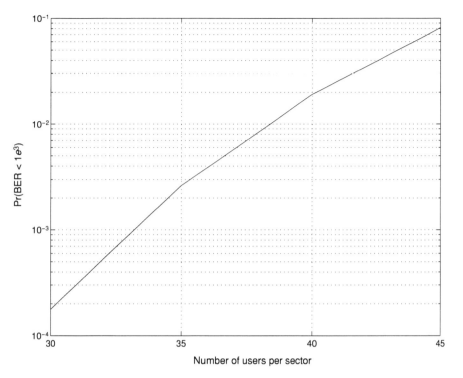

FIGURE 3.10: Downlink capacity.

the downlink supports a few more users (38) per sector than does the uplink. The limiting link appears to be the downlink, meaning that the number of users supported per sector is 36. If we assume a 5% blocking probability is desired, these 36 channels can support 30.7 Erlangs. Comparing this to the capacity determined by the CDMA Development Group, or CDG, (the commercial consortium of CDMA cellular providers), we find that this estimate is optimisitic. Specifically, the CDG quotes a capacity of 12–13 Erlangs for IS-95 (the second-generation standard) and 24–25 Erlangs for *cdma2000* (the third-generation cellular standard).

3.3.3   Capacity–Coverage Trade-Off

To this point, we have concerned ourselves with only the capacity of a CDMA system. However, because the system is interference-limited, a fundamental relationship—specifically, an inverse relationship—exists between the system capacity and the coverage area. Increasing the number of users in the system increases the uplink interference, which, if a target E_b / I_0 is to be maintained, requires an increase in the mobile transmit power. However, the mobile transmit power is limited, and thus the coverage area shrinks.

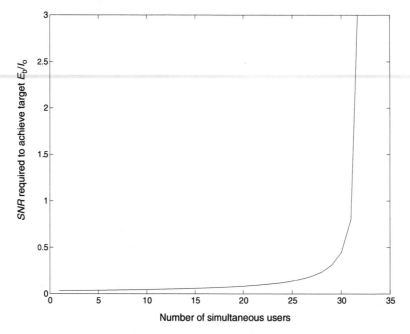

FIGURE 3.11: Illustration of pole capacity: The required SNR to maintain a target E_b/I_o grows exponentially with capacity.

This relationship can be clearly seen by taking (3.31), setting $I/P = 0$ (i.e., ignoring out-of-cell interference), and solving for SNR:

$$\frac{P}{N} = \frac{1}{(B_T/R_b)/(E_b/I_0) + 1 - K} \tag{3.48}$$

This function is plotted in Figure 3.11 for $R_b = 8$kbps, $E_b/I_0 = 7$dB, and $B_T = 1.25$MHz. Clearly, the required SNR grows dramatically with system loading. Again, since mobiles have limited transmit power, mobiles at farther distances will be unable to maintain the required SNR as the capacity grows and will thus fail to achieve the target E_b/I_0. This effectively reduces the coverage area or range of the cell. Also clear from Figure 3.11, the function seems to approach an asymptote as the number of users approaches 35. This is referred to as *pole capacity* K_{pole} and refers to the theoretical maximum number of users that can be supported. This value can be obtained by solving (3.31) for K and letting SNR approach infinity:

$$\begin{aligned} K_{pole} &= \lim_{P/N \to \infty} \left(\frac{B_T/R_b}{E_b/I_0} - \frac{N}{P} + 1 \right) \\ &= \frac{B_T/R_b}{E_b/I_0} + 1 \end{aligned} \tag{3.49}$$

For the parameters used in Figure 3.11, we have $K_{pole} = 32$, which agrees with the plot. Making the substitution for K_{pole} in (3.48), we arrive at an expression for SNR that explicitly shows the reason for the term *pole capacity*:

$$\frac{P}{N} = \frac{1}{K_{pole} - K}$$

(3.50)

where clearly the required SNR approaches infinity as K approaches K_{pole}.

3.3.4   Erlang Capacity

To this point, we have analyzed capacity in terms of the radio interface (or air-interface) capacity. In other words, we have examined the number of simultaneous signals that can be supported. However, in typical traffic analysis, we are interested in *Erlang capacity*, which reflects the fact that not all users use the system simultaneously but randomly access the system. This effect was discussed in Chapter 1 and impacts all types of cellular systems regardless of the multiple access technique. In cellular CDMA systems, there are two fundamental capacity limits: the air interface capacity limit and the hardware resource limit. For each CDMA channel being received at the base station, dedicated hardware must be available for demodulation, decoding, framing, and so on. Since dedicated hardware has an associated cost, minimizing the necessary hardware at the base station is a priority. However, sufficient channel resources (typically termed *channel elements* or CEs) must also be guaranteed to provide a required quality of service (i.e., blocking probability). While the air interface capacity is essentially limited on a sector-by-sector basis, CEs can be pooled across sectors, providing a cell-level trunking efficiency. Another factor that must be considered, soft hand-off also requires CEs and thus affects the Erlang capacity.

We can analyze the impact of channel pooling by considering the probability of blocking for different numbers of channel elements at the base station for 1–3 sectors. For a single sector system, assuming Poisson arrivals with rate of λ and a service time of $1/\mu$, the probability of blocking follows the Erlang B formula given in (1.15) and repeated here for convenience:

$$\Pr\{blocking\} = \frac{\frac{\Lambda^K}{K!}}{\sum_{k=0}^{K} \frac{\Lambda^k}{k!}}$$

(3.51)

where $\Lambda = \lambda/\mu$ and K represents the air interface limit (i.e., the total number of allowable channels). In this case, the number of CEs should be equal to the number of channels K. If it is smaller, the capacity is decreased directly since the air interface limit cannot be supported.

In the case of two sectors sharing a common pool of CEs, the performance is somewhat different. If the number of CEs is K_{CE}, then clearly if $K_{CE} \geq 2K$, the performance is identical to the single-sector case. However, we can reduce the total number of channel elements because we can allow the sectors to share channel elements. To see this, we follow Kim's development [3] and let $\mathbf{P}_A$ and $\mathbf{P}_B$ be the marginal probabilities for the two sector capacities in vector form. The ith entry of $\mathbf{P}_A$ represents the probability of i simultaneous users in sector A and is given by

$$\mathbf{P}_A(i) = \sum_{j=0}^{M} \mathbf{P}_{A|B}(i \,|\, j)\, \mathbf{P}_B(j) \tag{3.52}$$

where $\mathbf{P}_{A|B}(i \,|\, j)$ is the conditional probability of i users in sector A given j users in sector B. When $K_{CE} < 2K$, the probabilities are not independent since they must share a limited pool of CEs. In matrix form, we can write

$$\mathbf{P}_A = \mathbf{P}_{A|B}\mathbf{P}_B \tag{3.53}$$

The jth column of $\mathbf{P}_{A|B}$ is simply the state probability vector of sector A given that there are j users in sector B. When there are j users in sector B, there are only $K_{CE} - j$ CEs available for sector A. Thus, we can write

$$\mathbf{P}_{A|B}(i \,|\, j) = \begin{cases} \dfrac{\frac{\Lambda^i}{i!}}{\sum_{k=0}^{K}\frac{\Lambda^k}{k!}} & \begin{array}{l}\forall i \\ 0 \leq j \leq K_{CE} - K\end{array} \\[4ex] \dfrac{\frac{\Lambda^i}{i!}}{\sum_{k=0}^{K_{CE}-j}\frac{\Lambda^k}{k!}} & \begin{array}{l} 0 \leq i \leq K_{CE} - j \\ K_{CE} - K + 1 \leq j \leq K \end{array} \\[4ex] 0 & \text{otherwise} \end{cases} \tag{3.54}$$

If we assume that the Erlang load on the two sectors is the same, we can write

$$\mathbf{P}_A = \mathbf{P}_{A|B}\mathbf{P}_A \tag{3.55}$$

and we can solve for $\mathbf{P}_A$ as the eigenvector of $\mathbf{P}_{A|B}$ corresponding to an eigenvalue of one. Once $\mathbf{P}_A$ is obtained, we can obtain the joint probability matrix $\mathbf{P}_{AB}$ from

$$\mathbf{P}_{AB} = \left[\mathbf{P}_A(0) \begin{pmatrix} \mathbf{P}_{A|B}(0 \,|0) \\ \mathbf{P}_{A|B}(1 \,|0) \\ \vdots \\ \mathbf{P}_{A|B}(M|0) \end{pmatrix}, \mathbf{P}_A(1) \begin{pmatrix} \mathbf{P}_{A|B}(0 \,|1) \\ \mathbf{P}_{A|B}(1 \,|1) \\ \vdots \\ \mathbf{P}_{A|B}(M|1) \end{pmatrix} \cdots \right.$$

$$\left. \mathbf{P}_A(M) \begin{pmatrix} \mathbf{P}_{A|B}(0 \,|M) \\ \mathbf{P}_{A|B}(1 \,|M) \\ \vdots \\ 0 \end{pmatrix} \right] \tag{3.56}$$

Once we have determined the joint probability matrix, the blocking probability for sector A is simply equal to the probability that either K users are in sector A or K_{CE} users are in the two sectors combined:

$$\Pr\{blocking\} = \sum_{j=0}^{K_{CE}-K} \mathbf{P}_{AB}(K, j) + \sum_{j=K_{CE}-K+1}^{K} \mathbf{P}_{AB}(K_{CE} - j, j) \qquad (3.57)$$

where we have assumed that $K_{CE} < 2K$. The three-sector case can be found in a similar manner. Specifically, again assuming equal loading in each sector, one can write the sector A state probability vector as

$$\mathbf{P}_A = \mathbf{P}_{A|B+C}\mathbf{P}_{B+C} \qquad (3.58)$$

where $\mathbf{P}_{B+C}$ is the vector of marginal probabilities for the combined sectors B and C. Further, we can write

$$\mathbf{P}_{B+C} = \mathbf{P}_{B+C|A}\mathbf{P}_A \qquad (3.59)$$

and

$$\mathbf{P}_A = \mathbf{P}_{A|B+C}\mathbf{P}_{B+C|A}\mathbf{P}_A \qquad (3.60)$$

Thus, the marginal probability vector $\mathbf{P}_A$ is the eigenvector of the matrix $\mathbf{P}_{A|B+C}\mathbf{P}_{B+C|A}$ corresponding to an eigenvalue of 1. The conditional matrix $\mathbf{P}_{A|B+C}$ is found in the same manner as $\mathbf{P}_{A|B}$ in the two-sector case where the jth column represents the number of users in sectors B and C. The conditional matrix $\mathbf{P}_{B+C|A}$ is determined from the joint probability matrix from the two-sector case. Specifically, the columns of $\mathbf{P}_{B+C|A}$ can be written as

$$\mathbf{P}_{B+C|A}(\cdot\,|j) = \begin{bmatrix} P_{BC}(0, 0) \\ P_{BC}(0, 1) + P_{BC}(1, 0) \\ \vdots \\ \sum_{k=0}^{\min(i, N-j)} P_{BC}(k, i - k) \\ \vdots \\ \sum_{k=0}^{\min(2M, N-j)} P_{BC}(k, M - k) \end{bmatrix} \qquad (3.61)$$

Once $\mathbf{P}_A$ has been determined from (3.60), we can determine the joint probability matrix $\mathbf{P}_{B+C,A}$ from

$$\mathbf{P}_{B+C|A} = \left[\mathbf{P}_A(0) \begin{pmatrix} \mathbf{P}_{B+C|A}(0|0) \\ \mathbf{P}_{B+C|A}(1|0) \\ \vdots \\ \mathbf{P}_{B+C|A}(2M|0) \end{pmatrix}, \mathbf{P}_A(1) \begin{pmatrix} \mathbf{P}_{B+C|A}(0|1) \\ \mathbf{P}_{B+C|A}(1|1) \\ \vdots \\ \mathbf{P}_{B+C|A}(2M|1) \end{pmatrix} \cdots \right.$$
$$\left. \mathbf{P}_A(M) \begin{pmatrix} \mathbf{P}_{B+C|A}(0|M) \\ \mathbf{P}_{B+C|A}(1|M) \\ \vdots \\ 0 \end{pmatrix} \right] \tag{3.62}$$

The resulting probability of blocking can be written as

$$\Pr\{\text{blocking}\} = \sum_{j=0}^{K_{CE}-K} \mathbf{P}_{B+C,A}(j, M) + \sum_{j=K_{CE}-K+1}^{\min(2K,K_{CE})} \mathbf{P}_{B+C,A}(j, K_{CE}-j) \tag{3.63}$$

where the implicit assumption has been made that $K_{CE} < 3K$.

As an example, consider a single-sector cell with an air interface limit of $K = 10$. The blocking probability versus the number of CEs is plotted in Figure 3.12 for sector loads of $\Lambda \in \{2, 3, 4, 5\}$ Erlangs. The blocking probability drops until the $K_{CE} = K$. For a desired blocking probability of 1% and a load of $\Lambda = 3$ Erlangs, the system must have approximately eight CEs. For a load of $\Lambda = 4$ Erlangs, the system must have approximately nine CEs.

Now consider a two-sector system whose blocking probability is shown in Figure 3.13. If a load of $\Lambda = 3$ Erlangs is applied to each sector, a 1% blocking probability requires a total of $K_{CE} = 12$ CEs shared across the two sectors. At a load of $\Lambda = 4$ Erlangs per sector, approximately fifteen CEs are required. Notice that even though the overall load doubled, the number of required CEs did not double. Specifically, by sharing CEs, a savings of 25% (twelve versus sixteen CEs) is realized for $\Lambda = 3$ Erlangs.

Results for a three-sector system are plotted in Figure 3.14 for the same per sector loads. As in the two-sector case, pooling channels across three sectors provides a benefit: a reduction in the total number of required CEs. Specifically, if we again examine a load of 3 Erlangs per sector, a 1% blocking probability requires sixteen CEs. This is compared to eight CEs for a single-sector and twelve CEs for a two-sector system. Without channel pooling, a three-sector

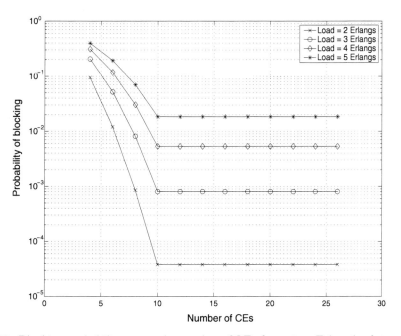

FIGURE 3.12: Blocking probability versus the number of CEs for various Erlang loads in a single sector ($K = 10$).

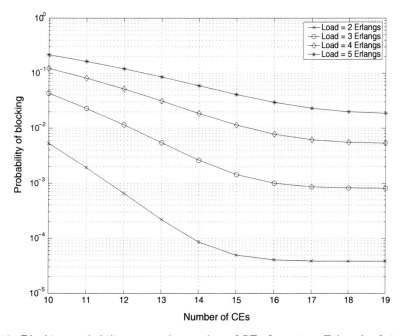

FIGURE 3.13: Blocking probability versus the number of CEs for various Erlang loads in a two-sector system ($K = 10$).

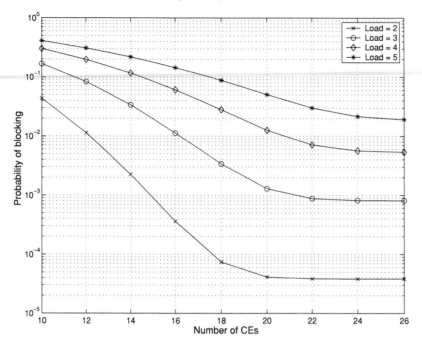

FIGURE 3.14: Blocking probability versus the number of CEs for various Erlang loads in a three-sector system ($K = 10$).

system would require 24 CEs. Thus, channel pooling allows a reduction of 33%. A similar gain can be observed at other loading factors shown in Figures 3.12 through 3.14.

3.4   RADIO RESOURCE MANAGEMENT

Radio resource management (RRM) is the process of optimizing the transmit power, spectrum, and channel allocation to maximize the number of users that can be provided a minimum quality of service for a specific set of base stations and coverage area. In CDMA systems, RRM is primarily a task of interference management. Interference management is accomplished through four basic functions: power control, base station assignment (hand-off), admission control, and load control. We will discuss each function in turn.

3.4.1   Power Control

In Section 2.6, we found that the capacity of a DS/CDMA system is sensitive to the relative received power of different signals due to the near-far problem. Ideally, we wish to have all signals received with nearly identical power.[1] To maintain this, we require *power control*, which

---

[1]The exact required receive power depends on the propagation characteristics.

can be designed to combat small-scale propagation effects (e.g., fast Rayleigh fading) or large-scale propagation effects (e.g., shadowing, system load) or both. Additionally, transmitter power control can rely on either feedback from the receiver or its own measurements. It is thus divided into two broad categories, open-loop and closed-loop, depending on whether feedback is used. Within closed-loop power control, there are often two complementary power control loops working at different rates. These two loops are termed the *inner loop* and the *outer loop*. It should be noted that while power control is a generic function, the taxonomy here has been heavily influenced by the original CDMA cellular standard, IS-95, which uses the terminology discussed. Additionally, for ease of discussion, we will use the uplink as the reference for the following discussion although the comments apply equally well to the downlink.

Open-Loop Power Control

The simplest form of power control is open-loop power control. In open-loop power control, the transceiver simply examines the received signal strength from the base station and adjusts its transmit power in an inverse manner, using received signal strength as an indicator of path loss. Such a scheme is fairly slow and can be inaccurate when multipath fading dominates performance because multipath fading effects on the uplink and downlink frequencies are generally uncorrelated [7]. However, if fast power control is unnecessary and multipath fading can be averaged out, the approach may provide sufficient performance.

Closed-Loop Power Control

A more sophisticated form of power control is closed-loop control in which the receiver feeds back either channel quality information or direct power control commands, depending on the amount of feedback that can be supported. Power control can be based on various channel quality indicators, but the most common is received signal strength. To easily understand this form of power control, consider a power control scheme similar to what is used in commercial CDMA systems. In IS-95, *cdma2000*, and UMTS, the base station receiver measures the received signal power, compares the value to a stored threshold, and sends back a single bit every T_{pc} seconds[2] to indicate whether the mobile should increase or decrease power by some fixed amount.

To understand the effect that power control has on system performance, let us consider three specific channel scenarios: an AWGN channel, a slow (relative to the power control update rate) fading channel, and a fast fading channel. Figure 3.15 shows the plots of the channel, transmit power, and received power for an AWGN channel with 500-Hz power control feedback. We assume that the receiver simply transmits a single bit indicating that the transmitter should increase or decrease the transmit power by 0.5dB and ignore measurement or feedback errors. The receiver threshold is assumed to be unity. The top plot in Figure 3.15

---

[2] T_{pc} = 1.25 ms in IS-95 and *cdma2000* and half that length in UMTS.

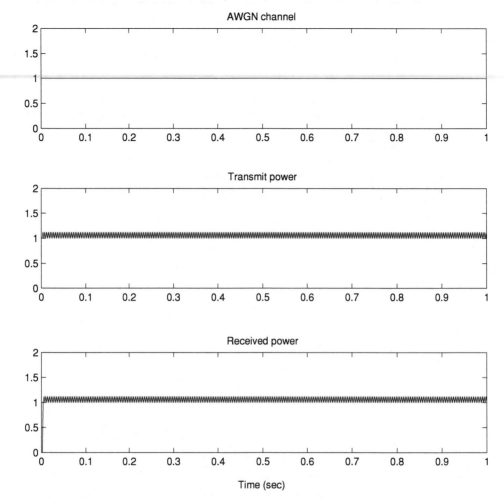

FIGURE 3.15: Illustration of power control in an AWGN channel.

shows the channel, which is assumed to be constant. The transmit power (and thus the received power) simply oscillates around the threshold value since the power control dictates that the power must increase or decrease every T_{pc} seconds. Fast power control is unnecessary in this case, but it does not hurt performance.

Now let us consider a slow fading channel (5 Hz Rayleigh fading) as shown in Figure 3.16. The top plot shows the envelope of the channel, which demonstrates that the channel varies slowly with time. Due to power control, the received power is nearly constant. This dramatically improves performance since it essentially converts a Rayleigh fading channel into an AWGN channel. However, while a dramatic improvement in the performance is seen from the receiver's viewpoint, the improvement is not free as can be seen from the transmitter's viewpoint. The

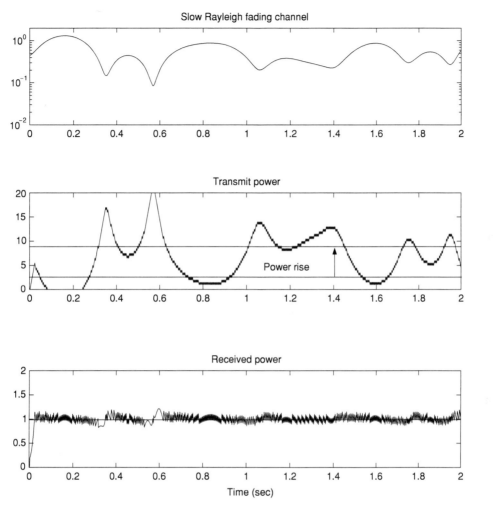

FIGURE 3.16: Illustration of power control in slow fading channel.

transmit power can be seen to essentially invert the fading channel. Because the transmitter must provide large increases in power to counteract deep fades, the average power is substantially increased. This is illustrated in the middle plot of Figure 3.16 and is termed a *power rise* at the transmitter. This increase in the average transmit power eats away at the overall system gain achieved by power control.

The last example that we wish to examine is a fast (150Hz) fading channel shown in Figure 3.17. The fading rate is substantially faster than what the power control can accommodate. Thus, the received signal varies wildly and power control is ineffective. In fact, it is worse than ineffective since, in addition to providing no benefit at the receiver, it also results in an increase in the average transmit power.

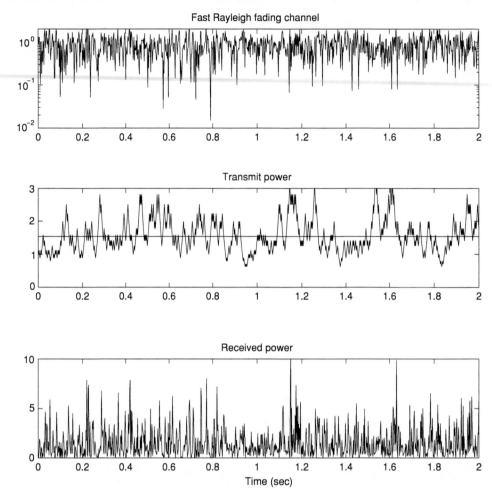

FIGURE 3.17: Illustration of power control in fast fading.

As described, closed-loop power control attempts to adjust the transmit power to achieve a target received signal power threshold. However, the needed threshold is dependent on system loading, fading statistics, the number of resolvable multipath components, and other factors. For this reason, the form of power control that we just described is often paired with a second closed-loop technique that controls the target threshold. In this two-loop arrangement, the loop that adjusts transmit power is termed *inner-loop power control* and a second loop that adjusts the received power threshold (sometimes called the E_b set-point) is called *outer-loop power control*.

An example of the outer loop is shown in Figure 3.18. In this example, the outer loop is controlled by frame errors. Specifically, the power control E_b set-point is changed every frame depending on whether the frame was in error. Through the use of check sums, the receiver can

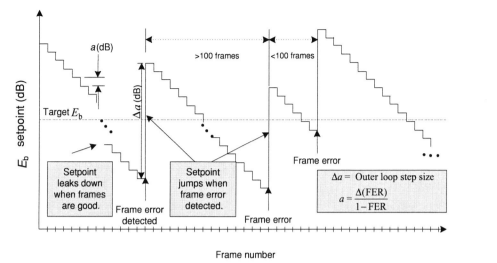

FIGURE 3.18: Illustration of outer-loop power control.

determine with high probability whether the frame was in error. If the frame was not in error, the set-point is reduced by $\Delta (FER)/(100 - FER)$dB where FER is a target frame error rate in percent. However, if the frame is in error, the set-point is increased by ΔdB. In steady state, the power will converge to a constant median power. Further, the set-point is reduced $100 - FER$ times per 100 frames and increased FER times, guaranteeing a frame error rate of FER. Thus, the outer loop provides a target performance by adjusting the inner-loop set-point as system parameters change.

3.4.2   Mobile-Assisted Soft Hand-Off

The second technique for radio resource management is base station assignment. Specifically, in CDMA systems, a mobile station is connected to multiple base stations simultaneously, which is termed *soft hand-off*. We introduced soft hand-off and described some of its benefits in Section 3.1.3. In this section, we provide more details of the technique. Due to universal frequency reuse, soft hand-off is natural in CDMA systems and requires no additional RF channels. An example procedure for hand-off is illustrated in Figure 3.19.

In a typical CDMA system, the mobile unit monitors the pilot signal of the base station to which it is currently communicating as well as the pilots from several other base stations. Each mobile has a list of base stations that are in its vicinity, termed its neighbor set. The mobile monitors the pilot strength of each pilot in its neighbor set. An example of the measured pilot strength during a hand-off is given in Figure 3.19. (1) When any pilot strength exceeds a threshold termed the *Add Threshold* (T_ADD), the mobile moves the pilot from its neighbor

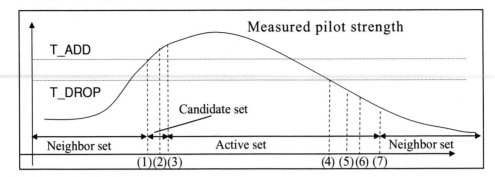

FIGURE 3.19: Illustration of soft hand-off.

set and into its candidate set. The mobile then requests a hand-off to that cell. (2) If the cell has sufficient resources, the mobile switching center will send a message to the base station and the mobile to begin a hand-off. (3) The mobile moves the pilot to its active set and completes hand-off. As long as the signal strength remains above a drop threshold (T_DROP), the signal will remain in its active set. The mobile then communicates simultaneously with all base stations in its active set. Most CDMA systems support at least three-way soft hand-off, with some supporting up to six-way soft hand-off. (4) When the pilot strength drops below the drop threshold, the mobile begins a hand-off drop timer. (5) When the hand-off drop timer expires, the mobile sends a hand-off message to the base station. (6) The base station then acknowledges receipt of the hand-off request by sending its own hand-off message. (7) Finally, the mobile terminates its connection and moves the pilot to its neighbor set.

Besides macro-diversity, soft hand-off ensures that a mobile is always communicating with the strongest base station in its view. In classic hard hand-off techniques, the hysteresis effect ensures that a mobile does not ping-pong between base stations. However, in doing so, the mobile is not always communicating with the strongest base station. This is tolerable, although not optimal, in FDMA/TDMA systems but is a problem in CDMA systems since it means that the strongest base station is actually causing substantial interference. Soft hand-off avoids this.

Finally, a distinction between soft hand-off between two base stations and soft hand-off between two sectors of the same base station must be explained. The latter is usually termed *softer hand-off*. Soft and softer hand-off look identical to the mobile station since it cannot distinguish between two cells and two sectors from the same cell. However, it makes a difference on uplink performance; in softer hand-off, uplink signals can be combined before decisions are made. In soft hand-off, separate decisions must be made on the uplink signals at the two base stations and decoded frames sent to the mobile switching center. However, softer hand-off typically fails to provide the same diversity advantage as soft hand-off.

3.4.3   Admission Control

Unlike TDMA/FDMA systems, CDMA systems have a soft capacity limit. That is, TDMA/FDMA systems have a specific number of channels available, and when they are all in use, the cell or sector is full. However, in CDMA, system capacity is determined predominantly by interference. Thus, the capacity limit is soft because it can always be broken provided a higher BER can be tolerated. Additionally, due to varying propagation conditions, the interference from a given number of mobiles can vary dramatically. Thus, there is no fixed limit on the number of users that can be supported.

Although the number of signals that can be supported is not fixed, a cell still cannot handle every request to enter the system. Determining whether or not to admit a new user is termed *admission control*. In CDMA, admission control cannot be based merely on the number of signals in the system but must be based on the amount of interference currently in the system and the amount of interference that a new user would generate. Typically, there are two separate load levels, which we will call Limit A and Limit B. Limit A is typically 60% of pole capacity for the uplink (or 60% of the transmit power for the downlink) and is the limit at which a base station (or sector) stops accepting new calls. Limit B is typically 85% of pole capacity (or 85% of transmit power for the downlink) and is the limit at which a base station stops accepting new calls and soft hand-off requests. By having a two-tier admission policy, a base station can control both the call blocking probability (the probability that a new call is unaccepted) and call drop probability (partly due to the soft hand-off failure).

To understand the admission control process, let us focus on the uplink. However, a similar analysis can clearly be done for the downlink. Systems engineers typically define a concept termed *system load*, which for CDMA systems can be defined as

$$\eta_{UL} = 1 - \frac{\sigma_n^2}{I_{total}} \qquad (3.64)$$

where

$$I_{total} = I_{ic} + I_{oc} + \sigma_n^2 \qquad (3.65)$$

σ_n^2 is the receiver thermal noise power, I_{ic} is the in-cell interference, and I_{oc} is the out-of-cell interference. Note that the load is a value between 0 and 1. Specifically,

$$\lim_{I_{total} \to \infty} \eta_{UL} = 1 \qquad (3.66a)$$

$$\lim_{(I_{ic} + I_{oc}) \to 0} \eta_{UL} = 0 \qquad (3.66b)$$

Now, rearranging (3.64), we can write the total interference as a function of the load:

$$I_{total} = \frac{\sigma_n^2}{1 - \eta_{UL}} \qquad (3.67)$$

We would like to know how the interference grows as the system load increases. Thus, we take the derivative of the interference with respect to the load:

$$\frac{d I_{total}}{d\eta_{UL}} = \frac{\sigma_n^2}{(1 - \eta_{UL})^2}$$

$$= \frac{I_{total}}{1 - \eta_{UL}} \qquad (3.68)$$

Thus, one way to estimate the interference increase due to a particular load increase is to use the approximation

$$\Delta I = \frac{I_{total}}{1 - \eta_{UL}} \Delta L \qquad (3.69)$$

where the increase in load is defined as

$$\Delta L = \frac{1}{1 + (B_T/R_b)/(v\, E_b/I_0)} \qquad (3.70)$$

Now let us look at a particular example with $B_T = 1.25\text{MHz}$, $R_b = 9.6\text{kbps}$, and $E_b/I_0 = 7\text{dB}$. The plot of interference versus load is given in Figure 3.20. When a new user requests access

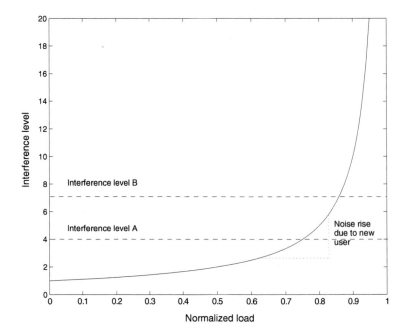

FIGURE 3.20: Illustration of admission control.

to the system, the admission control process examines the increase in cell interference at the current load. If the new interference level exceeds Level B, the request is denied. If the new interference level exceeds Level A, the request is denied if it is a new call but accepted if it is a soft hand-off request.

3.4.4   Load Control

In addition to admission control, CDMA systems must exercise load control to avoid amplifier overload on the downlink and excessive interference on the uplink. The most benign form of load control is to simply decrease the E_b/I_0 target at the base station for uplink control. This reduces the interference level seen while slightly degrading performance. This slight loss in performance is worth the additional stability afforded. On the downlink, load control can be accomplished by denying "up" power control commands from the mobile. More severe actions include amplifier overload control and dropping calls in a controlled fashion.

Amplifier overload control is a means for reducing the number of users in a cell by reducing the base station transmit power, particularly the pilot power. Mobile stations near the edge of coverage will automatically hand-off to surrounding cells since other pilots will now appear stronger than the current cell. This is illustrated in Figure 3.21. This phenomenon is also sometimes termed *cell-breathing*, which reflects that CDMA cells are not necessarily static. Cell sizes can be decreased by reducing the transmit pilot strength or by increasing the uplink interference level. As mentioned previously, as the system load increases, the cell size naturally shrinks since far away mobiles can no longer be adequately received.

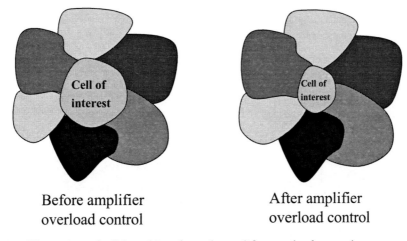

Before amplifier
overload control

After amplifier
overload control

FIGURE 3.21: Illustration of cell breathing through amplifier overload control.

3.5   SUMMARY

In this chapter, we have described the application of CDMA to cellular systems. Specifically, we showed that CDMA provides several positive properties including interference averaging, easy exploitation of voice activity, universal frequency reuse, and soft hand-off. These properties greatly enhance the overall capacity of cellular systems as compared to traditional TDMA or FDMA cellular systems. However, these same properties require sophisticated radio resource management techniques such as power control, mobile-assisted hand-off, load control, and admission control. These techniques are vital to CDMA since capacity is fundamentally connected to interference management on the uplink and transmit power management on the downlink.

CHAPTER 4

Spread Spectrum Packet
Radio Networks

In previous chapters, we have discussed the use of spread spectrum waveforms as a means of channelization in centralized wireless systems that are dominated by voice traffic. While this is the dominant use of CDMA in commercial systems, in military applications spread spectrum waveforms are also used in *distributed packet networks*. Such networks tend to use random access or other contention-based protocols for channel access. Spread spectrum can benefit such networks because of its resistance to multipath fading, ability to reject narrowband interference (e.g., jamming), low probability of detection or intercept, and enhanced multiple access capabilities. Additionally, in distributed packet networks, spread spectrum also offers an advantage over narrowband systems by providing the *capture effect*, which allows for successful reception in the presence of collisions under certain conditions (to be discussed later). It should be noted that when spread spectrum waveforms are used in such networks, the technique is typically referred to as spread spectrum multiple access (SSMA) as opposed to CDMA [1]. As the chapter title suggests, they are often referred to as *spread spectrum packet radio networks* (SS/PRNs) [43, 44].

The use of a spread spectrum based protocol for distributed packet radio networks was investigated at least as early as the 1980s [35, 44, 45]. Spread spectrum was proposed for packet radio networks due to its inherent capability to mitigate jamming and multipath fading in military applications. Unlike centralized systems where all uplink transmissions are multipoint-to-point and all downlink transmissions are point-to-multipoint, distributed networks have many point-to-point connections. Packet radio protocols are typically contention-based access techniques, such as ALOHA or CSMA as discussed in Chapter 1, due to the lack of centralized control.

There are three basic aspects of SS/PRNs: the spread spectrum radio protocol, the code assignment protocol, and the channel access technique. In terms of the spread spectrum protocol, SS/PRNs can be based on either direct sequence (Section 4.3), time-hopping, or frequency hopping (Section 4.4). First, we will discuss code assignment with DS/SS in some detail in the next section and briefly discuss channel access techniques in Section 4.2.

4.1   CODE ASSIGNMENT STRATEGIES

When spread spectrum is added to PRNs, several difficulties arise. Specifically, spread spectrum brings with it the possibility of multiple channels since multiple spreading codes are possible. With multiple channels, we now must determine which channels (i.e., codes) the receiver should monitor while in the idle state and on which code the node should transmit. Thus, in the SSMA context, a main difficulty with distributed networks is the assignment of spreading codes.

In terms of code assignment, there are three basic approaches: common code assignment, transmitter-based code assignment, and receiver-based code assignment. In the first approach, a single spreading code is used by all nodes in the system. Such a system is similar to traditional ALOHA or CSMA protocols with the exception that it is possible that multiple transmissions avoid annihilating each other if they are separated in time by more than a chip interval (i.e., the capture effect). However, if a Rake receiver is used with DS/SS, multiple transmissions will be difficult to separate. The original 802.11 protocol is an example of this type.

Two types of collisions occur in SS/PRNs: *primary collisions* and *secondary collisions*. Primary collisions occur whenever two users transmit on the same code at the same time. Secondary collisions occur whenever two users transmit on different codes at the same time. Primary collisions will typically result in packet errors whereas secondary collisions have the benefit of spreading gain to mitigate packet errors. Clearly, in a common code assignment approach, all collisions will be primary collisions.

The second possibility for code assignment is to assign all nodes an individual code for transmission, which is termed *transmitter-based assignment* [45]. Since each transmitter has a unique spreading code, multiple transmissions can occur simultaneously without packet annihilation, thus increasing system throughput. In fact, there will be no primary collisions (transmissions on the same spreading code) since all transmissions use different spreading codes by definition. The main difficulty with such an approach is that idle nodes do not know which code to monitor for incoming transmissions. Ideally, each receiver must monitor all spreading codes simultaneously, which is highly impractical with limited node complexity.

The third basic code assignment scheme is a receiver-based scheme where all nodes are assigned a specific code for receiving rather than transmitting. When node A has a packet to send to node B, it transmits the data on node B's spreading code. This eliminates the problem of receiver complexity since each node will listen to only its own spreading code. However, the downside is that primary collisions between transmissions can now occur since multiple transmissions on the same code are possible.

4.1.1   Common-Transmitter Protocol

We can solve some of the short-comings of these approaches by creating hybrid protocols, which combine features of the three approaches described above. Two specific hybrid protocols are the

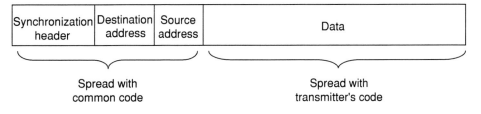

Synchronization header	Destination address	Source address	Data

Spread with common code   Spread with transmitter's code

FIGURE 4.1: Packet structure for C-T code assignment approach.

common-transmitter (C-T) protocol and the receiver-transmitter (R-T) protocol [44]. In the first method, a unique transmitting code is assigned to each user, and a common code is used for addressing purposes. For each transmission, the transmitter uses both the common code and its own unique transmitter code. Specifically, in the transmitted packet, the destination and the source addresses (along with a synchronization header) are transmitted first on the common code while the data is sent afterward on the transmitter's code (see Figure 4.1).

All idle receivers are initially listening to the common code, and, once they recognize their address, they shift to the transmitting station's code.

The only primary collisions that can happen in this scheme are during the header transmission when the synchronization sequence and addresses are being transmitted on the common code. Other transmissions can occur simultaneously since they will utilize different spreading codes. Of course, packet errors can occur due to secondary collisions if the number of collisions is sufficiently high or the relative powers are sufficiently different (i.e., the near-far problem).

An example of this network is given in Figure 4.2. In the example, four transmissions are occurring simultaneously. Node 1 is transmitting on the common code, Node 2 is transmitting on Code 2, Node 4 is transmitting on Code 4, and Node 7 is transmitting on the common code. Node 3 is listening on Code 4, and Node 6 is listening on Code 2. Since Node 5 is currently receiving no specific transmission, it is listening on the common code. Thus, there are secondary collisions at each of the receiving nodes since multiple transmissions are taking place. However, with sufficient spreading gain and power control, these collisions will not disrupt the other transmissions. On the other hand, a primary collision occurs at Node 5, which is listening to the common code. As a result, Nodes 1 and 7 will need to retransmit unless their signal is received with substantially more power than the other (the capture effect). If both the signals are received with substantially the same power, both may need to be retransmitted. However, if one signal dominates the total received signal, only the weaker of the two signals will need to retransmit. Additionally, if the two signals are received at substantially different times (much greater than one-chip duration), the receiver will typically capture the first arriving signal and reject the second. In this case, only that transmitter whose signal arrives second will need to retransmit.

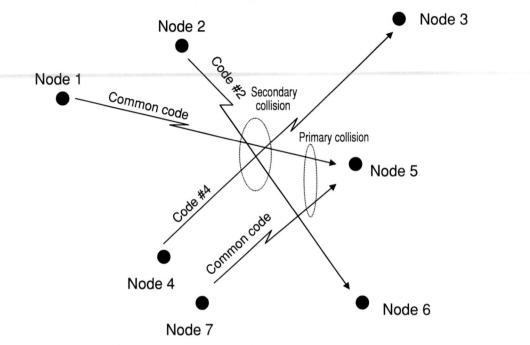

FIGURE 4.2: Illustration of C-T code assignment scheme for DS/SS/PRN.

A network based on the C-T code assignment protocol can be described using a state vector $\mathbf{s} = [m, n]$ where m is the number of communicating transmit/receive pairs and n is the number of transmitters whose transmissions are not being received. Consider a network of k nodes where the length of packet transmission is assumed to follow a geometric distribution. Using this description, one can show that the state transition probabilities (i.e., the probability of going from state $[k, l]$ to state $[m, n]$) can be written as [44]

$$p_{kl,mn} = q^{m+n-1} (1-q)^{k+l-m-n+1} p (1-p)^M *$$
$$\left\{ \binom{k}{m-1}\binom{l}{n} \frac{M^2+3M+2}{K-1} (1-p) - \right.$$
$$\left. \binom{k}{m}\binom{l}{n-1} \frac{M^2+M}{K-1} + \binom{k}{m} \sum_{i=1}^{l} \binom{l}{i} \binom{K-2m-i}{n-i} r^{n-i-1} \right\} \qquad (4.1)$$

where $M = K - 2m - n$, K is the total number of nodes, p is the packet transmission probability $r = p(1-q)/q$, and q is the parameter of the geometric distribution of the message length with an average message length $\overline{L} = 1/(1-q)$.

Example 4.1. Consider a system with a geometric distribution of packet length and an average packet length of $\overline{L} = 10$. What is the peak throughput (and at what packet transmission probability does it occur) for $K = 2$ users? Repeat for $K = 4, 8$, and 20.

Solution: The state transition probabilities can be determined from (4.1) and can be used to find the state probabilities via one of several well-known techniques. We find the eigenvector corresponding to the unit eigenvalue of the state transition matrix. If the state vectors are first converted to scalar values $k(m, n)$, the state transition probabilities can be represented as a matrix $\mathbf{P}$ where each element $P_{i,j}$ is the probability of transition from state i to state j. The state probabilities are then found as

$$\pi\mathbf{P} = \pi \qquad (4.2)$$

where π is the vector of state probabilities with $\pi_{k(m,n)}$ being the probability of being in state (m, n). The throughput is then found from

$$\kappa = \sum_{m,n} m\pi_{k(m,n)} \qquad (4.3)$$

since m is the number of successfully transmitting nodes. Figure 4.3 plots the throughput for packet transmission probabilities ranging from 0 to 1. As a point of comparison, if $\overline{L} = 1$, the maximum throughput approaches that of slotted ALOHA, e^{-1} packets per slot, as K gets large. However, for larger values of $\overline{L}$, the throughput increases since a smaller fraction of the

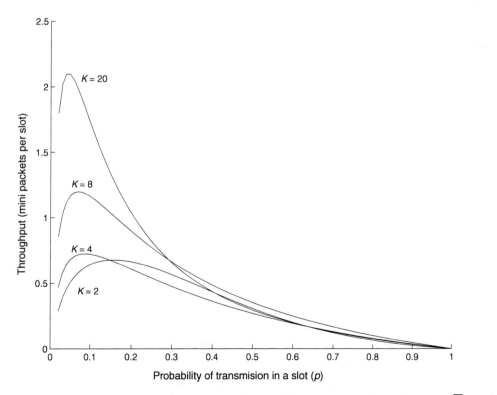

FIGURE 4.3: Throughput for the C-T protocol for SS/PRNs for various numbers of users, K ($\overline{L} = 10$).

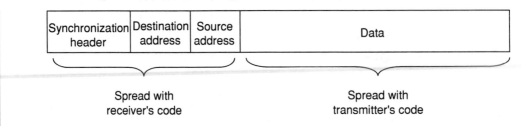

FIGURE 4.4: R-T code assignment protocol.

information is on the common code. The maximum throughput increases with K but occurs at $p < \frac{1}{K}$ which is a general feature for systems with $\overline{L} > 1$ [44]. As K increases, we also see that the throughput is more sensitive to the transmission probability due to the increased contention on the common code. The maximum throughput for $K = 2$ is 0.68, occurring at a transmission probability of approximately $p = 0.16$. Increasing the number of users to $K = 4$ increases the maximum throughput to 0.72 at a transmission probability of 0.085. As we increase the number of users to $K = 8$ and $K = 20$, the throughput increases to 1.2 and 2.1, respectively, but at transmission probabilities of 0.07 and 0.045.

4.1.2   Receiver-Transmitter Protocol

The R-T hybrid method assigns two spreading codes to every node. One of the codes is used for listening to incoming requests and the other code is used for transmitting data. Each transmission uses the intended receiver's receive code for spreading the synchronization header, the destination address, and the source address. The data is spread using the transmitter's transmit code. This is illustrated in Figure 4.4. The advantage of the R-T protocol over the C-T protocol is that there is a collision only when two (or more) header packets are received simultaneously at single destination node. The downside to the protocol is that multiple codes are required at each node. It also precludes the option of broadcasting information to the entire network.

These schemes are similar to the ALOHA protocol for narrowband systems. As such, improvements can be achieved by adding carrier sensing or reservation schemes. As an example, the protocols can be augmented by an RTS/CTS mechanism to avoid the loss of data [46]. The basic protocol remains the same except that there is a feedback loop in the setup stage. The adapted protocols are termed the multiple access collision avoidance/common-transmitter (MACA/C-T) and the multiple access collision avoidance/receiver-transmitter (MACA/R-T) protocols [46].

CSMA can also be added to the above techniques [47]. This is essentially a multi-channel protocol analogous to the MAC in the 802.11 standard for WLANs. In such a scheme, codes can be assigned dynamically for multiple access based on the communication needs of the node. The codes can be chosen from a set of pre-defined codes, and a particular node would choose a code that none of its adjacent nodes are currently using. The node can gather information regarding the codes being used in its vicinity by listening to a common channel where the initial communication setup is done. It can be shown that this protocol has tremendous improvement in performance compared to the CSMA and the RTS/CTS mechanism with a single spreading code because multiple sessions can occur simultaneously as adjacent nodes transmitting would ideally avoid interfering with each other.

Gerakoulis et al. [48] proposed an improvement over the receiver-based protocol [44], which gives the same throughput as the R-T–based scheme. Carrier sense is done at the transmitter on the receiver code of the intended receiver before it begins transmission, reducing the probability of collision. Contention is resolved using CSMA. Lo et al. proposed a similar scheme [49].

Almost all proposed CDMA-based MAC layers rely on static assignment of CDMA codes, which leads to inefficient utilization of the codes. In addition, as the number of required codes increases, the code length needs to be increased to obtain codes with good cross-correlation properties, which leads to a further reduction in throughput.

4.2   CHANNEL ACCESS STRATEGIES

In the centralized voice-centric systems described in Chapters 1–3, codes are assigned on a call-by-call basis, which is a classic CDMA approach. In packet radio systems, separate codes are not typically assigned on a call-by-call basis, so a channel access strategy is required as was discussed in the previous section. The channel access techniques are analogous to the random access methods described in Chapter 1. Specifically, SS/PRNs are typically designed to use techniques based on ALOHA, Slotted ALOHA, CSMA, and busy tone multiple access (BTMA) [43]. Note that unlike traditional narrowband random access protocols, channel access strategies for SS/PRNs allow for multiple simultaneous transmissions in the same geographic area.

The most common channel access technique used for SS/PRNs is ALOHA or Slotted ALOHA [43]. ALOHA protocols behave in a similar manner as ALOHA for narrowband techniques. Access to the channel is done in a random manner, and collisions are not resolved but simply require retransmission. The major difference with the use of spread spectrum is that multiple "channels" exist due to the possibility of multiple spreading codes being employed, depending on the code assignment strategy discussed earlier. For example, if receiver-oriented assignment strategies are employed, collisions occur when transmissions are directed to the same receiver but are avoided if the intended receivers are different. Additionally, due to the capture effect, multiple transmissions to the same receiver can result in one of the transmissions being successfully received.

As seen in Chapter 1, CSMA provides a substantial improvement in throughput over ALOHA techniques. However, ALOHA is still more common than is CSMA when spread spectrum is employed because channel sensing based on energy detection is impractical as an indicator of the success or failure of packet access since multiple transmissions are possible. Simply sensing that a transmission is taking place is inefficient since the transmitter needs to know if a particular spreading code is being used. To do this, acquisition is required, which adds complexity and delay to the sensing function. Additionally, if a transmitter-oriented protocol were used, a transmitter cannot know which spreading code to sense to determine if a transmission has occurred to a particular receiver.

Another option is the use of BTMA in which a subset of radios can hear that a given receiver is busy and transmits a busy tone. The reception of this busy tone suppresses transmission to the busy receiver. However, unlike traditional narrowband systems, the use of BTMA is meant not to prevent multiple transmissions but to simply prevent multiple transmissions to the same receiver.

4.3   DIRECT SEQUENCE PACKET RADIO NETWORKS

Like traditional CDMA systems, packet radio networks (PRNs) can be designed using either DS/SS or FH/SS. In this section, we will focus on direct sequence techniques. For a more detailed discussion of DS/SS, please refer to Chapter 2.

As discussed previously, the multiple access capabilities of DS/SS systems are dominated by code cross-correlation properties. This is also true of SS/PRNs. For centralized systems, power control also plays a large role in the multiple access capabilities due to the near-far problem. This problem is exacerbated in distributed networks due to the potential for an unintended receiver to be near a particular transmitter and has led many researchers to focus on frequency-hopped techniques for distributed systems.

In Chapter 2, we saw that the autocorrelation properties of the spreading codes in DS/SS systems determine the anti-multipath capabilities. However, in SS/PRNs, the autocorrelation properties also provide an additional benefit of the capture effect.

To understand this, consider a receiver-based code assignment strategy. If two transmitters want to transmit to a specific receiver, both will use the same spreading code when transmitting. Now let us assume that the distances between each of the transmitters and the intended receiver differ by more than the speed of light times the chip duration. In this case, a receiver tuned to the first incoming waveform will reject the second incoming waveform due to the good autocorrelation properties of the code used.

To better understand DS/SS-based SS/PRNs, consider the received signal at some node k. In any given slot, the node's receiver observes K active transmissions:

$$r(t) = \sum_{k=1}^{K} \sqrt{P_k} a_k(t - \tau_k) b_k(t - \tau_k) + n(t) \tag{4.4}$$

Assume that the C-T protocol is used and that code 1 is arbitrarily assumed to be the common code. Let us denote K_1 as the number of users transmitting on the common code for a particular frame period. The received signal can then be rewritten as

$$r(t) = \sum_{k=1}^{K_1} \sqrt{P_k} a_1 b_k (t - \tau_k) + \sum_{k=K_1+1}^{K} \sqrt{P_k} a_k b_k (t - \tau_k) + n(t) \qquad (4.5)$$

A node despreading using the common code observes the decision statistic

$$Z = \frac{1}{T} \int_0^T r(t) a_1(t) \, dt \qquad (4.6)$$

where we have aligned the timing with the desired signal's incoming signal for mathematical convenience ($\tau_1 = 0$). The despread statistic can then be written as

$$Z \simeq \sqrt{P_1} R_{11}(0) + \sum_{k=2}^{K_2} \sqrt{P_k} R_{11}(\tau_k) + \sum_{k=K_1+1}^{K} \sqrt{P_k} b_k R_{1k}(\tau_k) \qquad (4.7)$$

where we have ignored the data transmitted on the common code (assuming that the unmodulated common code is sent for acquisition) and the approximation is due to the symbol periods not exactly overlapping.

Consider a set of sequences that have mutual cross-correlation values of approximately $3/N$. Additionally, assume that the autocorrelation function is equal to $1/N$ for non-zero values. The decision statistic for the first arriving signal has a power gain of N^2 over the later-arriving paths and a gain of approximately $N^2/9$ over interfering signals. Thus, when decoding the first arriving signal, the cross-correlation and autocorrelation functions provide substantial isolation from primary and secondary collisions. First, since the autocorrelation function is small for non-zero offsets, primary collisions can be avoided provided that $\tau_k > T_c$ and the receiver does not employ a Rake architecture where it is looking for multipath components. Additionally, provided that the cross-correlation terms are small for all values of τ_k, secondary collisions can be rejected. Of course, both of these conclusions are based on the assumption that $P_1 \approx P_2 \approx P_k, \forall k$. If the received powers are dramatically different, the near-far effect is observed and the power disparity can more than offset the processing gain realized in the autocorrelation and cross-correlation functions. For this reason, DS/SS waveforms are not typically used for distributed networks.

Example 4.2.   Consider a system employing DS/SS with length 127 m-sequences with a chipping rate of 1Mcps and receiver-oriented code assignment. If two users transmit to a single receiver simultaneously, determine the SIR of the first user's despread signal. The first transmitter is 150m from the receiver, and the second transmitter is 750m from the receiver. Assume that power control is used such that signals arrive at the receiver with equal power.

Solution: The two signals are transmitted at the same time, and thus the second signal arrives $600\text{m}/(3*10^8 \text{ m/s}) = 2\,\mu\text{s}$. This corresponds to two chips. The autocorrelation function of an m-sequence is

$$R_{xx}[n] = \begin{cases} 1 & n = 0 \\ \frac{-1}{N} & n \neq 0 \end{cases} \tag{4.8}$$

Thus, for $n = 2$, the SIR is

$$SIR = \frac{PR_{11}^2[0]}{PR_{11}^2[2]}$$
$$SIR = N^2 \tag{4.9a}$$
$$= 127^2 \tag{4.9b}$$
$$= 42\text{dB}. \tag{4.9c}$$

4.4 FREQUENCY-HOPPED PACKET RADIO NETWORKS

Random access protocols for spread spectrum packet radio networks have also been investigated based on FH/SS [35,50–53]. The combination of coding and frequency hopping can be quite powerful as a random access technique. In fact, due to the near-far resistance of frequency hopping as compared to direct sequence, FH/SS protocols are more popular for distributed networks. Even with power control, distributed networks are prone to near-far problems due to the random locations of users.

Consider a distributed system illustrated in Figure 4.5. Transmit/receive pairs are uncoordinated and randomly hop among N frequencies, changing frequencies each symbol. Hopping each symbol is not strictly necessary, but hopping should occur at least several times per coded packet combined with pseudo-random interleaving. Assume that coding is used to correct

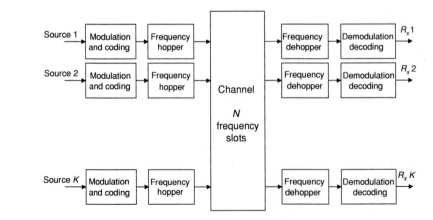

FIGURE 4.5: FHMA system with K transmit/receive pairs.

collisions, which are the dominant source of errors. In this case, the normalized throughput given K simultaneous users is [50, 53]

$$S(K) = \frac{r\,W(K)}{N} \qquad (4.10)$$

where r is the code rate used and $W(K)$ is the unnormalized throughput, which is equal to the number of users multiplied by the probability of a correct code word given K simultaneous transmissions:

$$W(K) = K * P_c(K) \qquad (4.11)$$

As discussed in Chapter 1, a common assumption in random access networks is that the number of active transmitters follows a Poisson distribution. That is, the probability of k active users is

$$p_K(k) = \frac{e^{-\lambda}\lambda^k}{k!} \qquad (4.12)$$

where λ is the average number of active users. The average normalized throughput can then be written as

$$\overline{S} = \sum_{k=0}^{\infty} p_K(k)S(k) \qquad (4.13a)$$

$$= \sum_{k=0}^{\infty} \frac{e^{-\lambda}\lambda^k}{k!} \frac{r\,k\,P_c(k)}{N} \qquad (4.13b)$$

$$= \frac{r}{N} \sum_{k=0}^{\infty} \frac{e^{-\lambda}\lambda^k}{k!} k\,P_c(k) \qquad (4.13c)$$

4.4.1   Perfect Side Information

To determine the average throughput, we require a specific code and its associated rate and error probability. Let us first assume that we are using a fixed-rate perfect code with rate r. In such a case, $P_c(k) = 1$, provided the rate does not exceed capacity. To determine capacity, we must consider the information available to the receiver. Specifically, it can be shown that the receiver can dramatically improve its performance if it knows when a collision or hit has occurred [1]. If the receiver knows perfectly when a hit has occurred, we label this case as *perfect side information*, and the receiver can label each hit as an erasure. Although this is not necessarily the optimal procedure, it provides a convenient framework for analysis [36]. Assuming M-ary frequency hopping, the channel with perfect side information can be modeled as an M-ary erasure channel that has transmission probabilities

$$P(y \mid x) = \begin{cases} 1 - p_{h,k} & y = x \\ p_{h,k} & y = \text{erasure} \\ 0 & \text{otherwise} \end{cases} \qquad (4.14)$$

where y is the received symbol, x is the transmitted symbol, and $p_{h,k}$ is the probability of a collision given k active users. The capacity of the channel is defined as the maximum mutual information transferred per channel use, and for an M-ary erasure channel with perfect side information and K users, this can be shown to be [54]

$$C_{PS}(K) = 1 - p_{h,K} \qquad (4.15)$$

Next, we must determine the probability of a hit given K active users. This probability depends on the hopping sequences and the presence or absence of synchronism between hopping sequences. Specifically, for memoryless independent hopping sequences, the probability of a hit for synchronous hopping can be shown to be [35]

$$p_{h,K} = 1 - \left(1 - \frac{1}{N}\right)^{K-1} \qquad (4.16)$$

This probability increases slightly for asynchronous hopping since two consecutive hops can collide with another user. Specifically, for the asynchronous case,

$$p_{h,K} \cong 1 - \left(1 - \frac{2}{N}\right)^{K-1} \qquad (4.17)$$

Finally, we can determine the average system throughput by choosing a code rate. Let us assume that the code rate is chosen to optimize throughput based on an average loading λ. The average throughput can then finally be written as [53]

$$\overline{S} = \frac{r_{opt}(\lambda)}{N} \sum_{k=0}^{\tilde{\lambda}} \frac{e^{-\lambda}\lambda^k}{k!} k \qquad (4.18)$$

where $\tilde{\lambda}$ is the limit on the number of users such that the capacity remains at or above $r_{opt}(\lambda)$. The probability of a correct code word is unity for $k \leq \tilde{\lambda}$ and zero otherwise.

A limitation of the above system is that the code rate is fixed regardless of the number of users currently in the system. If the code rate could be adapted to the number of active users,

$$\overline{S} = \frac{1}{N} \sum_{k=0}^{\tilde{\lambda}} \frac{e^{-\lambda} \lambda^k}{k!} k r_{opt}(k) \qquad (4.19a)$$

$$= \frac{1}{N} \sum_{k=0}^{\tilde{\lambda}} \frac{e^{-\lambda} \lambda^k}{k!} k C_{PS}(k) \qquad (4.19b)$$

Example 4.3. Consider a system with $N = 128$ frequency slots. What is the maximum throughput achievable with fixed rate coding and synchronous hopping? How does this compare to the maximum rate achievable with variable rate coding? Repeat for asynchronous hopping.

Solution: We must first determine the fixed code rate that maximizes throughput for synchronous hopping. This can be done via trial and error using (4.18) with $N = 128$. The final result is plotted in Figure 4.6. The throughput for variable rate coding is obtained using (4.19b)

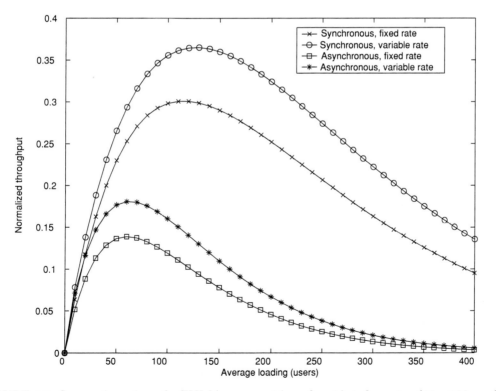

FIGURE 4.6: System throughput for FHMA system with perfect side information ($N = 128$, perfect codes).

and is plotted in Figure 4.6. From the plot, we can see that the maximum throughput for variable rate coding is approximately 0.36, and the fixed rate case obtains a maximum throughput of approximately 0.3. Thus, the variable rate coding provides a 20% improvement in maximum throughput. Repeating for the asynchronous case, we find similar results. The maximum throughput is less for the asynchronous case, but variable rate coding still provides a roughly 20% improvement. These trends are intuitive. The synchronous case provides better performance than does the asynchronous case as expected. This is analogous to the performance of ALOHA and Slotted ALOHA discussed in Chapter 1 (see Figure 1.8). Additionally, the use of fixed rate coding results in a throughput penalty since throughput is zero whenever the instantaneous loading level exceeds capacity.

4.4.2   No Side Information

In the opposite extreme in which no side information is available, the receiver fails to know when a collision occurs. In this case, we can model the channel as the M-ary symmetric channel with transmission probabilities

$$P(y \mid x) = \begin{cases} 1 - \left(\frac{M-1}{M}\right) p_{b,K} & x = y \\ \frac{p_{b,K}}{M} & x \neq y \end{cases} \tag{4.20}$$

where M is the source alphabet size. The channel capacity for the no side information case can be written as [53]

$$C_{NS}(K) = 1 + \left[\frac{M-1}{M} p_{b,K} \log_M \left(\frac{p_{b,K}}{M}\right) + \left(1 - \left(\frac{M-1}{M}\right) p_{b,K}\right) \right.$$
$$\left. \cdot \log_M \left(1 - \left(\frac{M-1}{M}\right) p_{b,K}\right) \right] \tag{4.21}$$

The average normalized system throughput is calculated as before with the exception that the capacity from (4.19b) is replaced by (4.21). As an example, consider Figure 4.7, which shows the throughput of synchronous and asynchronous systems with either fixed or variable rate coding for the same parameters as Example 4.3. Clearly, the throughput is reduced substantially without side information. Additionally, intuitive trends can be seen: variable rate coding does provides greater throughput than does fixed rate coding, and synchronous transmission provides greater throughput than does asynchronous transmission.

4.5   SUMMARY

In this chapter, we have described the application of spread spectrum waveforms to contention-based multiple access. Such multiple access techniques are useful in packet radio networks that cannot afford the overhead or infrastructure associated with scheduled access techniques but

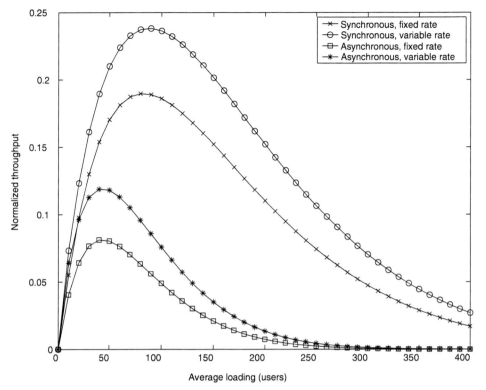

FIGURE 4.7: System throughput for FHMA system with no side information ($N = 128$, perfect codes).

wish to exploit the benefits of spread spectrum. Such networks can be based on either direct sequence or frequency hopping. In such networks, a primary consideration is the assignment of spreading waveforms, and we discussed the primary methods of code assignment. Due to the potential for severe near-far situations in ad hoc networks, frequency-hopped approaches are typically more appropriate. Thus, we focused on the network throughput for FHMA systems.

CHAPTER 5

Multiuser Detection

As discussed in previous chapters, the conventional matched filter receiver treats multiple access interference (MAI), which is inherent in CDMA, as if it were additive noise because, after despreading, the MAI tends toward a Gaussian distribution (see Chapter 2 for details). However, we have also seen that this MAI is the limiting factor in the capacity of CDMA systems. As a result, capacities for single-cell CDMA systems can be substantially lower than those for orthogonal multiple access techniques such as TDMA or FDMA. In addition, if one of the received signals is significantly stronger than the others, the stronger signal will substantially degrade the performance of the weaker signal in a conventional receiver due to the near-far problem. Thus, CDMA performance can be greatly enhanced by receivers designed to compensate for MAI. Multiuser receivers (sometimes referred to as *multiuser detection*) is one class of receivers that use the structure of MAI to improve link performance [55].

This chapter presents an overview of multiuser receivers and their usefulness for CDMA, particularly at the base station. As detailed in Chapter 3, cellular or personal communication system (PCS) design consists of two distinct problems: the design of the forward link (from the base station to the mobile) and the design of the reverse link (from the mobile to the base station). The forward link can be designed so that all signals transmitted to the mobiles are orthogonal and all signals arrive at the mobile receiver with similar power levels. Further, the mobile receiver must be inexpensive and have low power requirements. The reverse channel is potentially more harsh but can support a more sophisticated receiver. User signals arrive at the base station receiver asynchronously and can have significantly different energies, resulting in the near-far problem. In contrast to the mobile receiver, the base station receiver can be larger and more complex, have higher power consumption, and use information available about the interfering signals. We focus on this latter situation because it is more feasible that the receiver can simultaneously detect signals from all users (i.e., implement multiuser detection).

5.1 SYSTEM MODEL

To facilitate the discussion of the multiuser receiver structures presented in this chapter, we restate the model of DS-CDMA from Chapter 2. The received signal on the uplink can be represented as

$$r(t) = \sum_{k=1}^{K} s_k(t - \tau_k) + n(t) \qquad (5.1)$$

where K users are independently transmitting bi-phase modulated signals in an AWGN channel, τ_k is the delay of the kth user, $n(t)$ is a bandpass Gaussian noise process with double-sided power spectral density $N_0/2$, and

$$s_k(t) = \sqrt{2P_k} b_k(t) a_k(t) \cos(\omega_c t + \theta_k) \qquad (5.2)$$

where P_k and θ_k are the received power and phase of the kth user's signal, respectively, $b_k(t)$ is the data waveform, and $a_k(t)$ is the spreading waveform with spreading gain $N = T_b/T_c$. As discussed in Chapter 2, the uplink of a CDMA system is generally asynchronous. However, for ease of discussion, we will assume that signals are received synchronously ($\tau_1 = \tau_2 = \cdots = \tau_K = 0; \theta_1 = \theta_2 = \cdots = \theta_k = 0$) with random spreading codes. We arbitrarily examine the output of a filter matched to the kth user's spreading waveform during the first bit interval (assuming perfect carrier and PN code phase tracking). Assuming square symbol pulses, the correlator version of the matched filter is simply the integral of the received signal multiplied by the spreading code of interest, $a_k(t)$ and a phase-synchronous carrier $\cos(\omega_c t)$ over the symbol interval T_b

$$y_k = \frac{1}{T_b} \int_0^{T_b} r(t) a_k(t) \cos \omega_c t \, dt \qquad (5.3)$$

If this is repeated for each of K users, we can represent the set of matched filter outputs in vector notation as,

$$\mathbf{y} = \mathbf{RAb} + \mathbf{n} \qquad (5.4)$$

where $\mathbf{y} = [y_1, y_2, \ldots, y_K]^T$ and $\mathbf{R}$ is a $K \times K$ matrix that represents the correlation between spreading waveforms during the first bit interval. Thus, if $\rho_{j,k}$ are the elements of $\mathbf{R}$,

$$\rho_{j,k} = \frac{1}{T_b} \int_0^{T_b} a_j(t) a_k(t) dt \qquad (5.5)$$

$\mathbf{A}$ is a diagonal matrix with vector $[A_1, A_2, \ldots, A_K]^T$ along the diagonal, $A_k = \sqrt{P_k/2}$, $\mathbf{b} = [b_1, b_2, \ldots, b_K]$ are the data bits from each of the K signals, and $\mathbf{n} = [n_1, n_2, \ldots, n_K]^T$ is a vector of Gaussian noise samples with zero mean and covariance matrix $\Sigma_{\mathbf{n}} = \sigma^2 \mathbf{R}$. $\sigma^2 = N_o/4T_b$ is the noise power after dispreading. Decisions are then made as

$$\widehat{\mathbf{b}} = \text{sgn}(\mathbf{y}) \qquad (5.6)$$

where the function sgn($\mathbf{x}$) is applied element-by-element as

$$\text{sgn}(x) = \begin{cases} 1 & x \geq 0 \\ -1 & x < 0 \end{cases} \qquad (5.7)$$

As discussed in Chapter 2, the SINR Γ at the output of the matched filter depends on the number and relative power of the interferers. Specifically, the SINR for the kth signal is

$$\Gamma_k = \frac{P_k/2}{\dfrac{N_o}{4T_b} + \dfrac{1}{6N}\sum_{i \neq k} P_i}$$

$$= \left(\frac{N_o}{2E_b} + \frac{1}{3NP_k}\sum_{i \neq k} P_i \right)^{-1} \qquad (5.8)$$

Now this result assumed asynchronous reception with random phases between users. In general, it can be shown that the SINR of the kth signal is [32]

$$\Gamma_k = \left(\frac{N_o}{2E_b} + \frac{\varrho}{NP_k}\sum_{i \neq k} P_i \right)^{-1} \qquad (5.9)$$

where

$$\varrho = \begin{cases} 1 & \text{synchronous, zero phase} \\ 1/2 & \text{synchronous, random phase} \\ 2/3 & \text{asynchronous, zero phase} \\ 1/3 & \text{asynchronous, random phase} \end{cases} \qquad (5.10)$$

Now, assuming synchronous reception of signals with equal (zero) phase ($\varrho = 1$) and equal received powers for each signal ($P_i = P_k \, \forall i$) the SINR for each signal is equal and equal to

$$\Gamma = \left(\frac{N_o}{2E_b} + \frac{K-1}{N} \right)^{-1} \qquad (5.11)$$

Assuming Gaussian statistics, the probability of error of each signal is equal to

$$P_e = Q(\sqrt{\Gamma}) \qquad (5.12)$$

5.2 OPTIMAL MULTIUSER RECEPTION

It was widely believed for several years that because the MAI in a CDMA system tends toward a Gaussian distribution (i.e., because $\mathbf{y}$ is accurately modeled as a Gaussian random vector), the optimal receiver was the matched filter described above. However, since the MAI is in fact part of the desired signal, the optimal receiver is actually a joint detector that was first addressed by Schneider [56] for both the synchronous and asynchronous AWGN channels. Verdú expanded this work by more fully developing the mathematical model for the important case of the asynchronous channel and by determining the minimum receiver complexity [57]. Furthermore, Verdú developed probability of error bounds for the receiver.

For maximum likelihood sequence detection, we desire to maximize the joint *a posteriori* probability

$$P\left[\mathbf{b}\,|\,r(t)\right] \tag{5.13}$$

where $r(t)$ is the observed signal defined in (5.1). If all input vectors $\mathbf{b}$ are equally likely, this is equivalent to maximizing the *a priori* probability

$$P\left[(r(t)\,|\,\mathbf{b})\right] \tag{5.14}$$

For the AWGN channel, this maximization results in [55]

$$\hat{\mathbf{b}} = \underset{\mathbf{b}}{\mathrm{argmax}}[\exp\left(\Omega(\mathbf{b})/2\sigma^2\right)] \tag{5.15}$$

where $\Omega(\mathbf{b}) = 2\int_{-\infty}^{\infty} S(\mathbf{b})r(t)dt - \int_{-\infty}^{\infty} S^2(\mathbf{b})dt$, σ^2 is the noise power, and

$$S(\mathbf{b}) = \sum_k s_k(t, \mathbf{b}_k) \tag{5.16}$$

For the synchronous case, this is equivalent to finding the vector of bits $\mathbf{b}$ that maximizes [55]

$$p(\mathbf{y}\,|\,\mathbf{b}) = C\,\exp\left(-\frac{1}{2\sigma^2}(\mathbf{y} - \mathbf{RAb})^T\,\mathbf{R}^{-1}(\mathbf{y} - \mathbf{RAb})\right) \tag{5.17}$$

or

$$\hat{\mathbf{b}} = \underset{\mathbf{b}}{\mathrm{argmax}}\left[2\mathbf{b}^T\mathbf{Ay} - \mathbf{b}^T\mathbf{ARAb}\right] \tag{5.18}$$

Verdú showed that the gains of such a detector over the conventional matched filter receiver were dramatic [57]. We will examine the gains in the following example.

Example 5.1.   Consider a synchronous CDMA system with perfect power control ($\sqrt{2P_i} = 1$, $\forall i$), pseudo-random spreading codes, a spreading gain of $N = 15$, and $E_b/N_o = 7$dB. What is the performance advantage of the optimal receiver compared to the conventional matched filter in such a scenario?

Solution: The optimal receiver must search over 2^K binary sequences for the binary vector $\mathbf{b}$ that maximizes (5.17) for every bit interval. A closed form expression for the BER performance is not available and thus we must resort to simulation to determine the performance. On the other hand, the BER performance of the matched filter is well approximated by the Gaussian

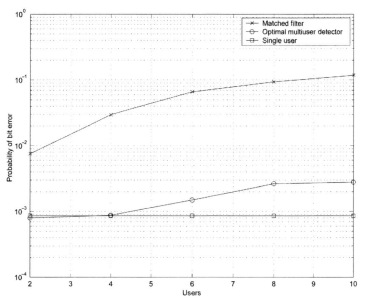

FIGURE 5.1: Comparison of the probability of bit error for the matched filter and optimal multiuser detector for a CDMA system (synchronous, random codes, $N = 15$, $E_b/N_o = 7$dB)

approximation as discussed in Chapter 2:

$$P_i = Q\left(\sqrt{\left(\frac{N_0}{2E_b} + \frac{K-1}{N}\right)^{-1}}\right) \qquad (5.19)$$

The BERs for the two detectors with the given parameters are plotted and compared to the single-user bound $P_e = Q\left(\sqrt{2E_b/N_0}\right)$ in Figure 5.1. We can see that the optimal detector provides dramatic performance improvement, nearly equaling the single-user bound. However, this performance improvement has come at a considerable computational complexity of $O(2^K)$ due to the search over all possible vectors $\mathbf{b}$. For asynchronous cases, the performance gain is equally dramatic although the complexity is even higher.

5.3   LINEAR SUB-OPTIMAL MULTIUSER RECEPTION

The previous section mentioned that while significant performance gains could be achieved over the conventional matched filter receiver, the cost of this performance gain is exponential complexity in the number of users. In this section, we investigate receivers that can approach the performance of the optimal receiver with significantly reduced computational complexity. These sub-optimal receivers can be broken down into two general categories, linear and non-linear, as shown in Figure 5.2. Linear sub-optimal receivers create data estimates based

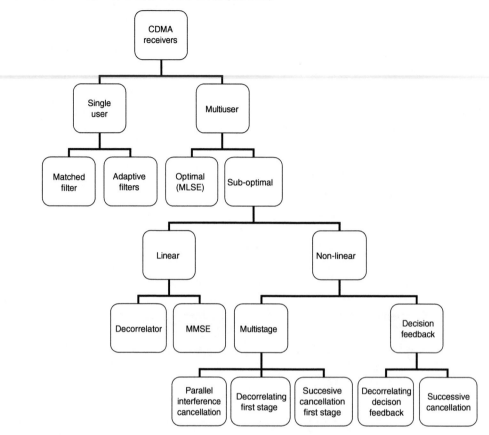

FIGURE 5.2: Receivers for CDMA systems

upon linear transformations of the sufficient statistics (i.e., the vector of matched filter outputs sampled at the symbol rate $\mathbf{y}$), and the non-linear implementations make decisions using non-linear transformations of the sufficient statistics.

5.3.1   The Decorrelating Detector

A linear detector is one that makes decisions based on a linear transformation of the matched filter output vector $\mathbf{y}$:

$$
\begin{aligned}
\hat{\mathbf{b}} &= \mathrm{sgn}(\mathbf{Ty}) \\
&= \mathrm{sgn}(\mathbf{T}(\mathbf{RAb} + \mathbf{n}))
\end{aligned}
\tag{5.20}
$$

where $\mathbf{T}$ is a linear operator on $\mathbf{y}$ and $\mathrm{sgn}(x)$ is defined in equation (5.7). This detector is illustrated in Figure 5.3 where the linear transformation is performed on the despread symbols and the matched filter is defined in equation (5.3). However, since dispreading is a linear operation, the linear transformation could obviously also be performed prior to dispreading or in conjunction with dispreading as we will show shortly.

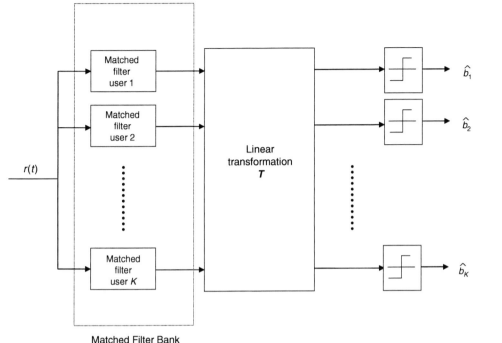

FIGURE 5.3: Block diagram of generic linear sub-optimal receiver structures

If the multiple access channel (excluding noise) is viewed as a deterministic multi-input multi-output linear filter with transfer function $\mathbf{R}$, then we can remove the interference in each of the matched filter outputs by applying the inverse transfer function. In other words, we can use the linear transformation $\mathbf{T} = \mathbf{R}^{-1}$, which leads to

$$\begin{aligned}\hat{\mathbf{b}} &= \mathrm{sgn}(\mathbf{R}^{-1}\mathbf{y}) \\ &= \mathrm{sgn}(\mathbf{Ab} + \mathbf{R}^{-1}\mathbf{n})\end{aligned} \qquad (5.21)$$

which is known as the decorrelating detector [58–60]. Since $\mathbf{R}$ is simply the normalized cross-correlation between the users' spreading codes, the decorrelating detector does not require knowledge of the received signal energies. In fact, the decorrelating detector is the optimal linear receiver when the signal energies are unknown [60]. This obviates the need for estimates of the received signal energies, which is a significant advantage since energy estimates tend to be extremely noisy.

Additionally, (5.21) shows that the data estimate of the kth user $\hat{b}_k$ is independent of the interfering powers. This can be seen from the fact that $\mathbf{A}$ is a diagonal matrix and eliminates the near-far problem discussed in Chapter 2. We will more fully discuss near-far resistance in Section 5.5.

There are, however, two main disadvantages of this receiver. The first is the need to calculate the inverse of the cross-correlation matrix $\mathbf{R}^{-1}$ to obtain the decorrelation coefficients. If the correlation matrix changes infrequently (i.e., the spreading codes change infrequently), this may not be a serious issue. However, if the matrix changes frequently or perhaps every symbol (as with long pseudo-random spreading codes), the complexity will be very high.

The second disadvantage is that in high noise situations (i.e., low E_b/N_o), the receiver performance can be severely degraded due to the enhancement of the noise power. In fact, the performance can actually be worse than that of the matched filter. More specifically, similar to the intersymbol interference (ISI) analog known as the zero-forcing equalizer, the application of the channel inverse results in increased noise power that is dependent on the cross-correlation between users. To show this, we examine the covariance matrix Σ_z of the decision metrics $\mathbf{z} = \mathbf{R}^{-1}\mathbf{y}$:

$$
\begin{aligned}
\sum\nolimits_{z} &= E\left[\mathbf{z}\mathbf{z}^T\right] - E\left[\mathbf{z}\right]E\left[\mathbf{z}^T\right] \\
&= E\left[\left(\mathbf{Ab} + \mathbf{R}^{-1}\mathbf{n}\right)\left(\mathbf{Ab} + \mathbf{R}^{-1}\mathbf{n}\right)^T\right] - (\mathbf{Ab})(\mathbf{Ab})^T \\
&= E\left[\mathbf{Ab}(\mathbf{Ab})^T + \mathbf{Ab}\left(\mathbf{R}^{-1}\mathbf{n}\right)^T + \mathbf{R}^{-1}\mathbf{nb}^T\mathbf{A}^T + \mathbf{R}^{-1}\mathbf{nn}^T\left(\mathbf{R}^{-1}\right)^T\right] - (\mathbf{Ab})(\mathbf{Ab})^T \\
&= \mathbf{Ab}E\left[\mathbf{n}^T\right]\left(\mathbf{R}^{-1}\right)^T + \mathbf{R}^{-1}E[\mathbf{n}]\mathbf{b}^T\mathbf{A}^T + \mathbf{R}^{-1}E[\mathbf{nn}^T]\left(\mathbf{R}^{-1}\right)^T \\
&= \sigma^2\left(\mathbf{R}^{-1}\right)^T
\end{aligned}
\tag{5.22}
$$

where we have used $E[\mathbf{n}] = 0$ and $E[\mathbf{nn}^T] = \sigma^2\mathbf{R}$ and σ^2 is the power of the AWGN at the output of the matched filter. Thus, the decorrelation process, while removing MAI, has also impacted the noise. We will examine the impact of this on the BER performance in a moment.

The decorrelating transformation can also be derived from the maximization of the likelihood function or, equivalently, the minimization of $(\mathbf{y} - \mathbf{Rb})^T\mathbf{R}^{-1}(\mathbf{y} - \mathbf{Rb})$ [61]. The probability of symbol error (equivalent to bit error in BPSK) of the kth user can be written as [60]

$$
P_e^k = Q\left(\sqrt{\frac{\left(E\left[z_k|b_k\right]\right)^2}{\text{var}\left[z_k\right]}}\right)
\tag{5.23}
$$

where z_k is the decision metric of the kth signal, $E\left[z_k|b_k\right] = A_k b_k$, var$[z_k]$ is the (k, k)th diagonal element of Σ_z, and $Q(\cdot)$ is the standard Q-function. Using these values and (5.22) in (5.23) results in

$$
\begin{aligned}
P_e^k &= Q\left(\sqrt{\frac{P_k/2}{(N_o/4T_b)\left(R^{-1}\right)_{k,k}}}\right) \\
&= Q\left(\sqrt{\frac{2E_b}{N_o}\frac{1}{\left(R^{-1}\right)_{k,k}}}\right)
\end{aligned}
\tag{5.24}
$$

where we have substituted for Σ^2 and N_0 is the one-sided noise power spectral density and we have assumed all data symbols are transmitted with equal probability. Thus, the performance of the decorrelator is identical to the single-user case with the exception of the noise enhancement factor $(\mathbf{R}^{-1})_{kk}$. Since all the elements of $\mathbf{R}$ are less than or equal to one, we find that $(\mathbf{R}^{-1})_{kk} > 1$. Unfortunately, general statistics of $\mathbf{R}^{-1}$ are not easily found, and thus predicting error probabilities is best done using the actual correlation matrix of a known set of user codes. One can obtain an estimate of the performance with random spreading codes by calculating the average of the elements along the diagonal of $\mathbf{R}^{-1}$ from simulations.

Example 5.2. Determine the BER performance of a two-user synchronous CDMA system with spreading codes that are the same every symbol interval when using a decorrelating detector receiver.

Solution: The correlation matrix for two users is written as

$$\mathbf{R} = \begin{bmatrix} 1 & \rho \\ \rho & 1 \end{bmatrix} \tag{5.25}$$

where $P = p_{12}$ defined in (5.5) and the inverse is then

$$\mathbf{R}^{-1} = \frac{1}{1 - \rho^2} \begin{bmatrix} 1 & -\rho \\ -\rho & 1 \end{bmatrix} \tag{5.26}$$

The resulting performance of both users is the same and determined from (5.24) as

$$P_b = Q\left(\sqrt{\frac{2E_b}{N_o} \left(1 - \rho^2\right)} \right) \tag{5.27}$$

Thus, we see that although the multi-access interference is eliminated, there is still a penalty paid. In fact, as the correlation between the two spreading codes increases, the performance degrades quickly.

The decorrelator as described requires no estimates of the users' received signal energies. However, it does require estimates of the timing and phase of each user along with knowledge of the spreading waveforms. Varanasi's work [62, 63] generalized the decorrelator to the case of noncoherent demodulation (specifically, differentially coherent phase shift keying) where neither the received amplitudes nor the received phases need to be estimated. This receiver, termed the bi-linear detector, simply performs the decorrelating operation followed by differential detection. The performance is again invariant to the interfering powers, thus obtaining near-far resistance. Additionally the receiver requires no phase tracking, assuming that the phase is constant over at least two consecutive symbol intervals.

The decorrelator can also be extended to non-AWGN channels. For example, the receiver can be used in flat and frequency selective Rician fading channels [64, 65] as well as flat and frequency selective Rayleigh fading channels (both coherent and differentially coherent) [66–71]. The most notable difference between these detectors and the AWGN case described earlier is that the multipath versions treat resolvable multipath as individual signals until a final decision is made. That is, the transform matrix **R** is now $KL \times KL$, incorporating each of the L paths from each user. The outputs for each path are then combined according to the desired diversity combination scheme (e.g., maximal ratio combining) after decorrelation. Further investigations have been published regarding the decorrelator in fading channels [64–71], temporal dispersion [72, 73], asynchronism [74–76], adaptive [77, 78] and reduced complexity [79–82]. Other variants of the decorrelating detector have also been reported [83–90].

Example 5.3. Show that the decorrelating detector can be formulated as a single-user detector and that the resulting receiver is simply a despreading operation with a modified spreading code.

Solution: The decision metric of the kth user is simply the kth element of the vector **z**

$$z_k = (\mathbf{R}^{-1}\mathbf{y})_k \qquad (5.28)$$

Defining $\langle \mathbf{a}, \mathbf{b} \rangle$ as the inner product of the vectors **a** and **b**, the matched filter output y_k can be written as the dot product of N samples of the filtered received signal and the spreading code $y_k = \langle \mathbf{r}, \mathbf{a}_k \rangle$ where r_k is kth chip-matched filter output

$$r_k = \frac{1}{T_b} \int r(t) p_{T_c}(t - kT_c) \cos(\omega_c t) dt \qquad (5.29)$$

and $\mathbf{a}_k$ is the vector of user k's spreading code values Thus, we can rewrite the decision metric as

$$z_k = \sum_{j=1}^{K} (\mathbf{R}^{-1})_{k,j} \, y_j$$

$$= \sum_{j=1}^{K} (\mathbf{R}^{-1})_{k,j} \langle \mathbf{r}, \mathbf{a}_j \rangle$$

$$= \left\langle \mathbf{r}, \sum_{j=1}^{K} (\mathbf{R}^{-1})_{k,j} \, \mathbf{a}_j \right\rangle$$

$$= \langle \mathbf{r}, \hat{\mathbf{a}}_k \rangle \qquad (5.30)$$

where $\hat{\mathbf{a}}_k = \sum_{j=1}^{K} (\mathbf{R}^{-1})_{k,j} \, \mathbf{a}_j$ is the modified spreading code of user k.

5.3.2   Linear Minimum Mean Squared Error Receiver

One of Verdú's key observations was the recognition of the analogous relationship between MAI cancellation and the equalization of ISI [91]. The decorrelator mentioned previously is very similar to the zero-forcing equalizer for ISI channels. This observation lead to other transfers of ISI techniques. As in ISI mitigation, linear detectors can be based on the minimum mean squared error (MMSE) criterion [61]. This detector attempts to minimize $E[(\mathbf{b} - \hat{\mathbf{b}})^T(\mathbf{b} - \hat{\mathbf{b}})]$ where $E[\cdot]$ is the expectation operation and $\hat{\mathbf{b}} = \text{sgn}(\mathbf{Ty})$ for some linear transform $\mathbf{T}$. The linear transformation $\mathbf{T} = \mathbf{M}$ that attains the minimum value is

$$\mathbf{M} = \left(\mathbf{R} + \sigma^2 \mathbf{A}^{-2}\right)^{-1} \qquad (5.31)$$

This leads to the decision rule

$$\hat{\mathbf{b}} = \text{sgn}\left[\left(\mathbf{R} + \sigma^2 \mathbf{A}^{-2}\right)^{-1} \mathbf{y}\right] \qquad (5.32)$$

It can be shown that the bit error rate of this receiver is upper-bounded by the bit error rate of the decorrelator [61]. Additionally, from (5.31) we can see that for $\sigma^2 \approx 0$ (i.e., high SNR situations), the MMSE detector is identical to the decorrelator.

$$\mathbf{M}_{high} \approx \mathbf{R}^{-1} \qquad (5.33)$$

In the opposite extreme (σ^2 is extremely large), the MMSE detector reduces to the conventional receiver. That is, for extremely low SNR values,

$$\mathbf{M}_{low} \approx \frac{2E_b}{N_o}\mathbf{I} \qquad (5.34)$$

where we have assumed that all values of A_k are equal. If they are not equal, the transformation matrix is still (approximately) diagonal and thus the transformation simply applies a constant scaling factor to the matched filter outputs which does not impact the decision.

Like the decorrelating receiver, the MMSE receiver obtains optimal near-far resistance; however, unlike the decorrelator, it requires knowledge of the received user powers. While some performance improvement is found (especially in the high noise case), the cost can be high. However, this detector lends itself well to adaptive implementation [92]. Specifically, if training sequences are employed, knowledge of all users' spreading codes is unnecessary. Instead, adaptive updates of the tap weights can converge to the necessary coefficients. The concept of adaptive MMSE and weighted least squares (WLS) based detectors (which account for MAI as well as ISI channel effects) is developed in several works [73, 93, 94].

5.4   NON-LINEAR SUB-OPTIMAL RECEIVERS: DECISION FEEDBACK

The previous section limited the discussion of multiuser detectors to linear sub-optimal detectors. As in the case of equalization, larger performance improvements can often be realized if non-linear techniques are applied, and a classical example of non-linear equalization is decision feedback [22]. As with equalization, the concept of decision feedback can be applied to multiuser channels. The concept of using decisions is sometimes termed *interference cancellation* due to the use of explicit regeneration and cancellation of multi-access interference. In this section, we will discuss four basic types of non-linear detectors: decorrelating decision feedback detectors, successive interference cancellation (SIC) receivers, parallel interference cancellation (PIC) receivers and general multi-stage receivers.

5.4.1   Decorrelating Decision Feedback

It was noted previously that a main drawback of the decorrelating receiver was the noise enhancement introduced by forcing the MAI to zero. In high noise situations, this can significantly degrade receiver performance. This problem can be avoided through a factorization of the cross-correlation matrix. This approach leads to a decorrelating decision feedback receiver [97]. Specifically, considering a synchronous channel, the cross-correlation matrix $\mathbf{R}$ (since it is positive definite) can be factored as

$$\mathbf{R} = \mathbf{F}^T \mathbf{F} \qquad (5.35)$$

where $\mathbf{F}$ is a lower triangular matrix. If a filter with response $(\mathbf{F}^T)^{-1} = \mathbf{F}^{-T}$ is then applied to the matched filter outputs, we obtain from (5.4) and (5.20)

$$\begin{aligned} \hat{\mathbf{y}} &= \mathbf{F}^{-T}\mathbf{y} \\ &= \mathbf{FAb} + \hat{\mathbf{n}} \end{aligned} \qquad (5.36)$$

where $\hat{\mathbf{n}} = \mathbf{F}^{-T}\mathbf{n}$ is a white Gaussian noise vector with covariance matrix $\sum_{\hat{\mathbf{n}}} = \sigma^2 \mathbf{I}$. In this transformation, we have whitened the noise, but we have not totally decorrelated the matched filter outputs. Since $\mathbf{F}$ is lower triangular, we have eliminated all the MAI from the first signal's decision statistic. Additionally, we have eliminated all the MAI from the second signal's decision statistic except the interference due to signal 1. The third signal's decision statistic contains interference only from signals 1 and 2, and so on. If the received signals are ranked according to their signal strengths, we can utilize our most confident decision metric first and ideally suffer the least from noise enhancement. Additionally, if we then subtract the estimate of the detected signal from the total received signal, detection of the second user signal can be improved. Assuming proper estimation, we can successively subtract users from the received signal until all users are detected. In this way, previous decisions are "fed back" to assist in the current decision. The detector for the strongest user is equivalent to the decorrelator. Assuming correct decisions, the weakest user's detector approaches the single-user bound. This factorization can be viewed as a combination of an equalization feed-forward filter $[(F^T)^{-1}]$, which eliminates multiuser interference due to future symbols and a feedback filter based on interference cancellation. The optimal feedback filter

$$\mathbf{D} = \mathbf{F} - \text{diag}(\mathbf{F}) \qquad (5.37)$$

operates on previously made bit decisions where $\text{diag}(\mathbf{X})$ is a diagonal matrix containing the diagonal of $\mathbf{X}$. $\mathbf{D}$ is thus a lower triangular matrix with zeros along its diagonal. The feedback terms at the output of this filter eliminate all MAI provided that feedback data is correct. After feedback, the input to the decision devices is given by

$$\mathbf{z} = \bar{\mathbf{y}} - \mathbf{DA}\hat{\mathbf{b}} \qquad (5.38)$$

and $\hat{\mathbf{b}} = \text{sgn}[\mathbf{z}]$. This formulation results in a bit estimate

$$\hat{\mathbf{b}} = \text{sgn}[\text{diag}(\mathbf{F})\mathbf{Ab} + \mathbf{n} + \boldsymbol{\xi}] \qquad (5.39)$$

where $\boldsymbol{\xi}$ is the residual interference due to incorrect past decisions, i.e., $\boldsymbol{\xi} = \mathbf{DA}(\mathbf{b} - \hat{\mathbf{b}})$. It can be shown that the feed-forward and feedback filters are optimal in terms of SNR [98]. As with the decorrelator, a drawback of the decorrelating decision feedback detector is the need to calculate the matrix coefficients, perform a matrix inversion, and in this case, perform a Cholesky decomposition. One method to avoid this calculation is to perform adaptive decorrelation [85].

The probability of error for the decision feedback, assuming that the feedback terms are correct can be shown to be [97]

$$P_e^k = Q\left(\sqrt{\frac{2E_b}{N_o}\mathbf{F}_{k,k}^2}\right) \qquad (5.40)$$

It can be shown that $\mathbf{F}_{k,k}^2 \geq 1/\mathbf{R}_{k,k}^{-1}$ where equality is obtained for signal 1. Thus, for the strongest received signal, the decorrelator and the decision feedback receivers present the same probability of error. For the other signals, the performance of the decision feedback strategy is superior to that of the decorrelator provided feedback estimates are correct. For the weakest user, the ideal performance of decision feedback matches the single-user bound since $F_{K,K} = 1$. One implementation issue for this receiver structure is the estimation of the received energies. This estimation error will cause some degradation in the performance of the receiver.

Example 5.4. Compare the performance of the matched filter, decorrelating detector, MMSE detector, and decorrelating decision feedback detector for a two-user synchronous system with fixed spreading codes as the cross-correlation between codes ρ varies between 0 and 0.95.

Solution: For equally likely bits, $1/2$ of the time, the bits of the two users reinforce each other, while $1/2$ of the time the bits negate each other. Thus, the matched filter performance for either user can be written as

$$P_e^{conv} = 0.5 Q\left(\sqrt{\frac{2E_b}{N_o}(1-\rho)^2}\right) + 0.5 Q\left(\sqrt{\frac{2E_b}{N_o}(1+\rho)^2}\right) \tag{5.41}$$

The performance of the decorrelator was determined in Example 5.2 as

$$P_e^d = Q\left(\sqrt{\frac{2E_b}{N_o}(1-\rho^2)}\right) \tag{5.42}$$

The performance of the MMSE receiver can be shown to be [55]

$$P_e = Q\left(\sqrt{\frac{2E_b}{N_o}\frac{(\mathbf{MR})_{11}^2}{(\mathbf{MRM})_{11}}\frac{1}{1+\lambda^2}}\right) \tag{5.43}$$

where

$$\lambda^2 = \frac{2E_b}{N_o}\frac{(\mathbf{MR})_{11}^2}{(\mathbf{MRM})_{11}}\left(\frac{A_2(\mathbf{MR})_{12}}{A_1(\mathbf{MR})_{11}}\right)^2 \tag{5.44}$$

The decorrelating decision feedback detector will attain the performance of the decorrelator for user 1 and the probability of bit error for signal 2 (assuming $\rho < 0.5$) is

$$P_e = \left(1 - P_e^d\right) Q\left(\sqrt{\frac{2E_b}{N_o}}\right) + P_e^d\left[0.5 Q\left(\sqrt{\frac{2E_b}{N_o}(1-2\rho)^2}\right) + 0.5 Q\left(\sqrt{\frac{2E_b}{N_o}(1+2\rho)^2}\right)\right] \tag{5.45}$$

for where P_e^d is the probability of error for the decorrelator. If $\rho > 0.5$, the second Q-function term should be replaced by $1/2$.

The performance plot for each receiver is given in Figure 5.4. We can see that as ρ increases, the matched filter degrades quickly. The decorrelating detector provides strong benefits for moderate values of ρ but provides diminishing benefits as the cross-correlation grows. For the two-user case with reasonable SNR, the benefit of using the MMSE receiver or decision feedback is not substantial. However, in heavier loading situations this may not always be the case. The decision feedback detector is particularly limited in the two-user case since the performance of the first decoded signal will be identical to the decorrelating detector.

5.4.2 Successive Interference Cancellation

A similar but somewhat simpler receiver structure using decision feedback is the successive interference canceller, shown in Figure 5.5 [99]. In this method, all users are again ordered in

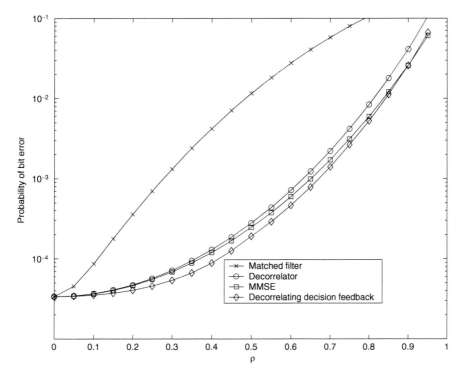

FIGURE 5.4: Impact of ρ on performance of various detectors in the two-user example

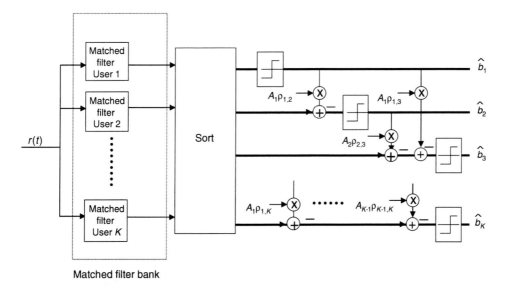

FIGURE 5.5: Block diagram of SIC (Despreading prior to cancellation)

decreasing received signal power (other ordering criteria are also possible). However, in this detector structure, the linear transformation described for the decorrelating decision feedback detector is eliminated. Rather, the matched filter outputs are used directly to make decisions. Typically, the strongest signal (i.e., the signal with the largest received power) is detected first. The decision made on this signal is then fed back to the detector of the second signal to improve the estimate of that signal. The previous decisions are utilized, along with an estimate of the signals' received energies and knowledge of the spreading codes, to remove the effects of the previous bits. In other words, this detector is similar to the decorrelating decision feedback receiver without the feed-forward filter. Formally, the bit estimate can be written in terms of the matched filter outputs as

$$\hat{b}_k = \operatorname{sgn}\left[y_k - \sum_{i=1}^{k-1} A_i \hat{b}_i \rho_{i,k} \right] \tag{5.46}$$

where we have assumed perfect channel knowledge (i.e., perfect knowledge of A_i). Of course, in practice, the channel must be estimated.

This formulation assumes the implementation of cancellation directly on the matched filter outputs as shown in Figure 5.5. However, cancellation can also be performed on the signal prior to dispreading (either before or after demodulation). In this case we have

$$\hat{b}_k = \operatorname{sgn}\left[\frac{1}{T_b} \int_0^{T_b} \left[r(t) - \sum_{i=1}^{k-1} 2A_i \hat{b}_i a_i(t) \cos(\omega_c t) \right] a_k(t) \cos(\omega_c t) dt \right] \tag{5.47}$$

As mentioned, the signal with estimates that are the most reliable (reliability can be estimated on the basis of various criteria [100, 101]) is detected first for several reasons. First, since the strongest users have the most reliable estimates, we can be more confident about the quality of the signal after cancellation. Additionally, the strongest signals are more robust to interference, thus requiring less cancellation for proper detection. Finally, the strongest signal causes the most interference and thus canceling it provided the most benefit.

The previous formulation of SIC is clearly non-linear due to the decision feedback. A non-trivial issue in this receiver structure is the estimation of the amplitudes A_i which are required for cancellation. The quality of these estimates can dominate the performance of the technique. Outside power estimates can be avoided by using the output of conventional correlators (i.e., the matched filter output) as a *combined* estimate of the signal amplitude and bit [102, 103]. This information can also be used to rank the signals for cancellation. The correlator output is simply multiplied by the user spreading code to regenerate the interference. This soft-decision approach eliminates the need for separate amplitude estimates. If the matched filter outputs are used as combined estimates of the transmitted bit and the channel gain in this fashion

[103–105], SIC is linear. This can be seen be examining the decision metrics. The decision metric of the kth user is

$$z_k = \frac{1}{T_b} \int_0^{T_b} \hat{r}_k(t) a_k(t) \cos(\omega_c t) dt \qquad (5.48)$$

where $\hat{r}_k(t)$ is the kth user's received signal after signals 0 through $k-1$ have been estimated and cancelled:

$$\hat{r}_k(t) = r(t) - \sum_{i=1}^{k-1} 2 y_i a_i(t) \cos(\omega_c t) \qquad (5.49)$$

Using this linear form of SIC, we can write the SINR for the kth detected signal using a Gaussian approximation for the MAI. Specifically, we can write the SINR Γ_k of signal k as [102]

$$\Gamma_k = \left\{ \left[\frac{\sigma^2}{P_k} + \frac{\varrho}{N} \sum_{i=2}^{K} \frac{P_i}{P_k} \right] \left(1 + \frac{\varrho}{N} \right)^{k-1} - \frac{\varrho}{N} \sum_{i=2}^{k} \left(1 + \frac{\varrho}{N} \right)^{k-i} \frac{P_i}{P_k} \right\}^{-1} \qquad (5.50)$$

where for rectangular chips ∂ is defined in (5.10) and $\sigma^2 = N_o/2T_b$ is the variance of the noise at the output of the baseband matched filter (note that factor of 2 difference from the bandpass variance). For the synchronous situation assumed throughout this chapter $\varrho = 1$.

Example 5.5.   One drawback of the SIC receiver is that, for equal received powers, the performance varies dramatically from signal to signal. To obtain equal BER performance, the received powers must follow a geometrically increasing profile. Determine the required received signal power profile for a non-linear SIC receiver (assuming perfect cancellation) for a desired *SINR* of 6dB after dispreading when long pseudo-random spreading codes are used, $N = 64$, and $K = 40$ users. Compare that to the required received profile for linear SIC. Assume that the signals are received synchronously with random phases in an AWGN channel.

Solution: In order to obtain equal BER, assuming an AWGN channel, all signals must obtain equal SINR. Assuming perfect cancellation, the SINR of the Kth decoded user at the output of the matched filter is simply

$$\Gamma = \frac{P_K}{\sigma^2} \qquad (5.51)$$

where $\sigma^2 = N_o/2T_b$ is the noise power at the output of the despreading filter. The required received signal power is for the Kth user is thus $P_K = \Gamma \sigma^2$. Under a perfect cancellation assumption and assuming random spreading codes, the signal of the $(K-1)$th decoded user is only affected by the Kth user's signal, thus the average SINR is

$$\Gamma = \frac{P_{K-1}}{\sigma^2 + \frac{\varrho}{N}P_K} \tag{5.52}$$

The required received signal power is found as

$$\begin{aligned} P_{K-1} &= \Gamma\left(\sigma^2 + \frac{\varrho}{N}P_K\right) \\ &= P_K\left(1 + \frac{\varrho}{N}\Gamma\right) \end{aligned} \tag{5.53}$$

The received SINR signal $K - 2$ is similarly written as

$$\Gamma = \frac{P_{K-2}}{\sigma^2 + \frac{\varrho}{N}(P_K + P_{K-1})} \tag{5.54}$$

Solving for the required received signal power for signal $K - 2$ results in

$$\begin{aligned} P_{K-2} &= \Gamma\left(\sigma_n^2 + \frac{\varrho}{N}(P_K + P_{K-1})\right) \\ &= \frac{P_K}{\sigma_n^2}\left(\sigma_n^2 + \frac{\varrho}{N}\left(P_K + P_K\left(1 + \frac{\varrho}{N}\Gamma\right)\right)\right) \\ &= P_K\left(1 + \frac{\varrho}{N}\Gamma + \frac{\varrho}{N}\Gamma\left(1 + \frac{\varrho}{N}\Gamma\right)\right) \\ &= P_K\left(1 + \frac{\varrho}{N}\Gamma\right)^2 \end{aligned} \tag{5.55}$$

Now, writing the SINR of signal $K - 3$ and solving for P_{K-3} results in

$$\begin{aligned} P_{K-3} &= \frac{P_K}{\sigma_n^2}\left(\sigma_n^2 + \frac{\varrho}{N}(P_K + P_{K-1} + P_{K-2})\right) \\ &= \frac{P_K}{\sigma_n^2}\left(\sigma_n^2 + \frac{\varrho}{N}\left(P_K + P_K\left(1 + \frac{\varrho}{N}\Gamma\right) + P_K\left(1 + \frac{\varrho}{N}\Gamma\right)^2\right)\right) \\ &= P_K\left(\left(1 + \frac{\varrho}{N}\Gamma\right)^2 + \frac{\varrho}{N}\Gamma\left(1 + \frac{\varrho}{N}\Gamma\right)^2\right) \\ &= P_K\left(1 + \frac{\varrho}{N}\Gamma\right)^3 \end{aligned} \tag{5.56}$$

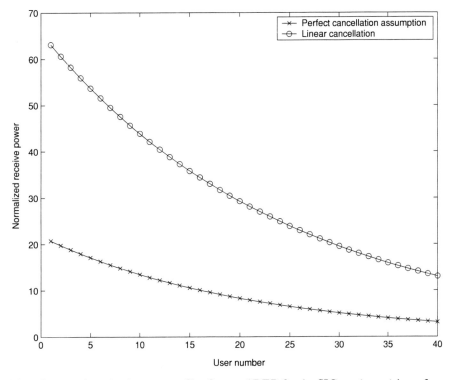

FIGURE 5.6: Required received power profiles for equal BER for the SIC receiver with perfect nonlinear cancellation and linear cancellation (AWGN, $\Gamma = 5$dB, $N = 64$, 40 users, synchronous, and random phase)

One can show that in general, the required power of the kth signal (where numbering is based on cancellation order) is [99]

$$P_k = P_K \left(1 + \Gamma \frac{\varrho}{N}\right)^{K-k} \tag{5.57}$$

and again $P_K = \sigma^2 \Gamma$ where Γ is the desired SINR for all signals [99]. For $\Gamma = 5$dB, $N = 64$ and $\varrho = 1/2$, the resulting profile is plotted in Figure 5.6.

Assuming linear cancellation, we can use the formulation for SINR in (5.50). Inverting the formula and multiplying through by P_K results in

$$\frac{1}{\Gamma_k} P_k = \sigma^2 \left(1 + \frac{\varrho}{N}\right)^{k-1} + \frac{\varrho}{N}\mathbf{x}_k^T\mathbf{p} \tag{5.58}$$

where $\mathbf{p} = [P_1, P_2, \ldots P_K]$ and

$$\mathbf{X}_{k,m} = \begin{cases} 0 & m = 1 \\ \left(1 + \dfrac{\varrho}{N}\right)^{k-1} - \left(1 + \dfrac{\varrho}{N}\right)^{k-m} & m \leq k \\ \left(1 + \dfrac{\varrho}{N}\right)^{k-1} & m > k \end{cases} \qquad (5.59)$$

Putting the power equation (5.58) into vector notation and assuming equal SINR values, i.e., $\Gamma_k = \Gamma, \forall k$ we have

$$\frac{1}{\Gamma}\mathbf{p} = \sigma^2 \beta + \frac{\varrho}{N}\mathbf{X}\mathbf{p} \qquad (5.60)$$

where β is a $K \times 1$ vector with $\left(1 + \frac{\partial}{N}\right)^{i-1}$ as the ith element and $\mathbf{X}$ is a $K \times K$ matrix with x_i^T as the ith row. Finally, solving for $\mathbf{p}$ we find that the following received power profile is required [102]:

$$\mathbf{p} = \left(\frac{1}{\Gamma}\mathbf{I} - \frac{\varrho}{N}\mathbf{X}\right)^{-1} \beta \sigma^2 \qquad (5.61)$$

where $\mathbf{I}$ is a $K \times K$ identity matrix.

The resulting required received powers are plotted in Figure 5.6 for both cases. Clearly, the perfect cancellation assumption leads to a lower bound (i.e., an optimistic estimate) on the required received signal powers for non-linear SIC, whereas linear SIC provides a more pessimistic (and realistic) view of the required powers. However, it is clear that the received signal powers increase geometrically in either case. In fact, the power ratio for perfect non-linear cancellation is

$$\frac{P_{k+1}}{P_k} = \left(1 + \frac{\varrho}{N}\Gamma\right) \qquad (5.62a)$$

and

$$\frac{P_{k+1}}{P_k} = \frac{\left(1 + \dfrac{\varrho}{N}\Gamma\right)}{\left(1 + \dfrac{\varrho}{N}\right)} \qquad (5.62b)$$

for linear SIC. This can be accomplished naturally in systems with widely varying received signal powers such as cellular systems along with FER-based power control [103, 106, 107].

5.4.3   Parallel Interference Cancellation

The main drawback of the SIC receiver is that, in the presence of equal received powers, the BER performance is poor for the signals that are detected early. In fact, the SINR for the first detected signal is equivalent to the matched filter. This can be seen explicitly for linear SIC from equation (5.50) with $k = 1$ and $\varrho = 1$:

$$
\begin{aligned}
\Gamma &= \left\{ \left[\frac{\sigma^2}{P_1} + \frac{\varrho}{N} \sum_{i=1}^{K} \frac{P_i}{P_1} \right] \left(1 + \frac{\varrho}{N} \right)^{1-1} - \frac{\varrho}{N} \sum_{i=2}^{1} \left(1 + \frac{\varrho}{N} \right)^{1-i} \frac{P_i}{P_1} \right\}^{-1} \\
&= \left\{ \left[\frac{\sigma^2}{P_1} + \frac{\varrho}{N} \sum_{i=1}^{K} \frac{P_i}{P_1} \right] \right\}^{-1} \\
&= \left\{ \left[\frac{N_o}{2E_b} + \frac{K-1}{N} \right] \right\}^{-1}
\end{aligned}
\tag{5.63}
$$

where we have used $\sigma^2 = N_o/2T_b$ and the result is the same as the matched filter given in (5.11).

A third type of non-linear receiver structure, the PIC receiver, alleviates this problem by providing equal cancellation benefit to all signals [58, 108–110]. This structure is plotted in Figure 5.7. PIC detectors use matched filters to estimate the data from all signals in parallel. The estimates for each user can then be used to reduce the interference to and from the other signals by subtracting the estimate of each interferer from the desired user's signal. Ideally, this would allow the elimination of all interference from the desired user. Formally,

$$
\hat{b}_k = \text{sgn} \left[y_k - \sum_{i \neq k} A_i \hat{b}_i \rho_{i,k} \right]
\tag{5.64}
$$

where again we have assumed perfect channel knowledge (i.e., A_i). Of course, in practice, this must also be estimated. Additionally, during this development we have assumed equal phase between users for notational simplicity. However, in practice there are clearly phase differences between users. This must also be estimated and used in the cancellation process. In such a case, we can consider A_i to be complex containing both amplitude and phase. Further, the final decision statistic would have to be phase rotated prior to making a decision.

The above formulation assumes the implementation of cancellation directly on the matched filter outputs (sometimes referred to as a narrowband implementation). Since cancellation and despreading are linear operations, we can perform cancellation prior to despreading with no change in performance. If cancellation is performed on the signal prior to despreading (sometimes termed a wideband implementation), we have

$$
\hat{b}_k = \text{sgn} \left[\frac{1}{T_b} \int_0^{T_b} \left[r(t) - \sum_{i \neq K} 2A_i \hat{b}_i a_i(t) \cos(\omega_c t) \right] a_k(t) \cos(\omega_c t) dt \right]
\tag{5.65}
$$

which is demonstrated in complex baseband form (i.e., after demodulation) in Figure 5.7.

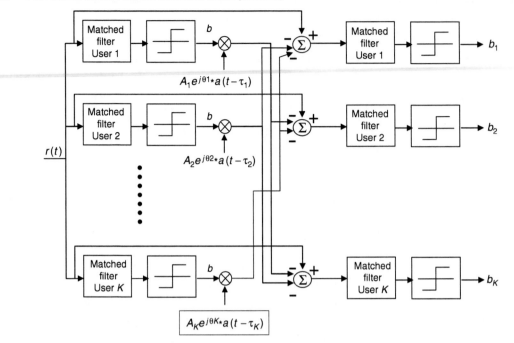

FIGURE 5.7: Parallel Interference Cancellation (non-linear, wideband cancellation implementation)

As in SIC detectors, PIC can be implemented in linear form. Specifically, we can directly use the matched filter outputs without making hard decisions as a combined estimate of the data bit and channel gain:

$$\hat{b}_k = \text{sgn}\left[\frac{1}{T_b} \int_0^{T_b} \left[r(t) - \sum_{i \neq K} 2y_i a_i(t) \cos(\omega_c t) \right] a_k(t) \cos(\omega_c t) dt \right] \qquad (5.66)$$

The analytical performance of this parallel cancellation approach in an AWGN channel can be determined using the standard Gaussian approximation for MAI. The resulting bit error rate of the receiver for the kth signal can be shown to be [111, 108]

$$P_e^k = Q\left(\left[\frac{N_o}{2E_b} \left(\frac{1 - \left(\frac{K-1}{N}\right)^2}{1 - \left(\frac{K-1}{N}\right)} \right) + \frac{1}{N^2} \left(\frac{(K-1)^2 - 1}{K} \frac{\sum_{j=1}^{K} P_j}{P_k} + 1 \right) \right]^{-1/2} \right) \qquad (5.67)$$

where K is the number of users and N is the processing gain [111]. Note that for asynchronous transmission N must be replaced by $3N$. The development of this equation assumes that y_k is an unbiased estimate of the $A_k b_k$. Unfortunately, it is found that this is not the case [112]. Rather, y_k is biased after cancellation with the bias increasing with system loading. One method of alleviating this problem is to multiply the estimate by a partial cancellation factor with a value in the range [0, 1] [112]. We will discuss partial cancellation shortly.

5.4.4   Multistage Receivers

The final non-linear multiuser receiver that will be discussed is the *general multistage receiver*, so termed because whenever decisions are made, they can be used either to make a final decision on the data or to enhance the signal through cancellation, which leads to another stage of detection as shown in Figure 5.8. As an example, consider the PIC detector examined in the last section. There is no reason why the bit estimates defined by (5.64) couldn't be used to perform a second round (i.e, stage) of cancellation before making a decision. In fact, cancellation could be performed an arbitrary number of times before a final decision is made. More specifically, the bit estimates defined in (5.64) can be used iteratively:

$$\hat{b}_k^{(s)} = \text{sgn}\left[y_k - \sum_{i \neq k} A_i \hat{b}_i^{(s-1)} \rho_{i,k} \right] \qquad (5.68)$$

where $\hat{b}_k^{(0)}$ is determined from the original matched filter outputs.

In this multi-stage receiver, a PIC detector is used at each stage. In general, any scheme could be used at each stage. For example, a decorrelating detector could be used in the first stage followed by multiple stages of parallel interference cancellation [113]. This would improve the initial estimates allowing for fewer stages to obtain a specific level of performance. Multiple stages of SIC detection could also be used [114]. However, the most popular form of multistage receiver is the multistage PIC receiver [108].

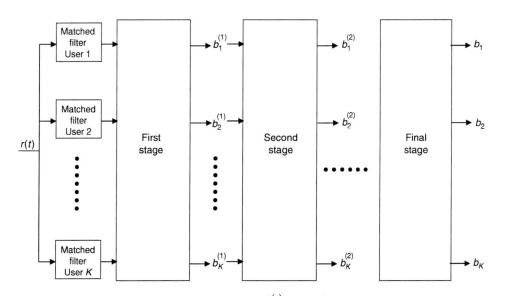

FIGURE 5.8: Illustration of the multistage detector ($\hat{b}_k^{(s)}$ is the bit estimate of the kth user after the sth stage of cancellation)

Mathematically, we can represent the bit decisions for an S-stage parallel cancellation scheme at any stage s as

$$\hat{b}_k^{(s)} = \text{sgn}\left[z_k^{(s)}\right] \tag{5.69}$$

where $z_k^{(s)}$ is the decision metric after s stages of cancellation:

$$z_k^{(s)} = \frac{1}{T_b} \int_0^{T_b} \hat{r}_k^{(s)}(t) a_k(t) \cos(\omega_c t) dt \tag{5.70}$$

and for $s > 0$, $\hat{r}_k^{(s)}(t)$ is the kth user's signal after s stages of cancellation:

$$\hat{r}_k^{(s)}(t) = r(t) - \sum_{i \neq k} 2A_i \hat{b}_i^{(s-1)} a_i(t) \cos(\omega_c t) \tag{5.71}$$

where i represents the index summing over all interfering signals. Again, the cancellation can be done before or after despreading. Further, assuming a matched filter in the first stage $\hat{r}_k^{(0)}(t) = r(t)$ and the decision statistics at stage 0 are [110]

$$\begin{aligned} \mathbf{z}^{(0)} &= \mathbf{y} \\ &= \mathbf{RAb} + \mathbf{n} \end{aligned} \tag{5.72}$$

which is equivalent to the matched filter outputs given in (5.4). Finally, putting together the previous equations, the bit estimate for signal k estimated at stage s, $\hat{b}_k^{(s)}$, is

$$\hat{b}_k^{(s)}(t) = \text{sgn}\left[y_k - \sum_{i \neq k} A_i \rho_{ik} \hat{b}_k^{(s-1)}\right] \tag{5.73}$$

The BER performance of the general multistage receiver is difficult to determine. However, if the intermediate stages are linear, the performance of the linear multistage PIC receiver with S stages of cancellation can be approximated as [111, 108]:

$$P_e^{(S)} = Q\left(\left[\frac{N_o}{2E_b}\left(\frac{1 - \left(\frac{K-1}{N}\right)^S}{1 - \left(\frac{K-1}{N}\right)}\right) + \frac{1}{N^S}\left(\frac{(K-1)^S - (-1)^S}{K}\frac{\sum_{j=1}^K P_j}{P_k} + (-1)^S\right)\right]^{-1/2}\right) \tag{5.74}$$

Note that for $S = 0$ stages of cancellation, the performance collapses to (5.19) as expected.

Although the BER performance of the general multistage reciever is difficult to determine analytically, we can gain some insight into the performance but examining the two–user situation with one stage of cancellation stages (termed a two–stage receiver). Let us consider the BER performance of a two–stage receiver with parallel cancellation in the second stage and a conventional first stage as compared to a two–stage receiver with a decorrelating first stage.

Assuming a conventional matched filter first stage, the bit estimate for signal 1 at the output of the second stage can be written as

$$\begin{aligned} \hat{b}_1 &= \text{sgn}\left[y_1 - \rho\text{sgn}(y_2)\right] \\ &= \text{sgn}\left[b_1 + \rho b_2 - \rho\text{sgn}(b_2 + \rho b_1 + n_2) + n_1\right] \end{aligned} \tag{5.75}$$

The bit estimate for signal two is similar. Determining the bit error rate for this case is not straightforward due to the correlation between n_1 and n_2. Specifically, due to symmetry the probability of bit error can be written as

$$P_e = \frac{1}{2}\left\{\Pr\left(\hat{b}_1 = -1\,|\,b_1 = 1, b_2 = 1\right)\right\} + \frac{1}{2}\left\{\Pr\left(\hat{b}_1 = -1\,|\,b_1 = 1, b_2 = -1\right)\right\} \quad (5.76)$$

Unfortunately, even conditioned on b_1 and b_2, since n_1 and n_2 are not independent, the BER must be determined by integrating the over the joint distribution of n_1 and n_2. Thus, we must rely on simulation or numerical integration.

Equation (5.76) also holds for the performance with a decorrelating first stage. However, in this case because of the decorrelating operation, the noise terms are independent and the problem can be solved in closed form. Specifically,

$$\begin{aligned}
\hat{b}_1 &= \text{sgn}\left[y_1 - \rho\text{sgn}(y_2 - \rho y_1)\right] \\
&= \text{sgn}\left[b_1 - \rho b_2 - \rho\text{sgn}(b_2 - \rho b_1 + n_2 - \rho(b_1 + \rho b_2 + n_1)) + n_1\right] \\
&= \text{sgn}\left[b_1 - \rho b_2 - \rho\text{sgn}(b_2(1 - \rho^2) + n_2 - \rho n_1) + n_1\right] \quad (5.77)
\end{aligned}$$

Now, since n_1 and $n_2 - \rho n_1$ are independent Gaussian random variables, we can write

$$P_b = \frac{1}{2}\left\{\Pr\left(\hat{b}_1 = -1\,|\,b_1 = 1, b_2 = 1, \text{sgn}(y_2) = -1\right)\Pr\left(\text{sgn}(y_2) = -1\,|\,b_1 = 1, b_2 = 1\right)\right\} + \ldots$$

$$\frac{1}{2}\left\{\Pr\left(\hat{b}_1 = -1\,|\,b_1 = 1, b_2 = 1, \text{sgn}(y_2) = 1\right)\Pr\left(\text{sgn}(y_2) = 1\,|\,b_1 = 1, b_2 = 1\right)\right\} + \ldots$$

$$\frac{1}{2}\left\{\Pr\left(\hat{b}_1 = -1\,|\,b_1 = 1, b_2 = -1, \text{sgn}(y_2) = -1\right)\Pr\left(\text{sgn}(y_2) = -1\,|\,b_1 = 1, b_2 = -1\right)\right\} + \ldots$$

$$\frac{1}{2}\left\{\Pr\left(\hat{b}_1 = -1\,|\,b_1 = 1, b_2 = -1, \text{sgn}(y_2) = 1\right)\Pr\left(\text{sgn}(y_2) = 1\,|\,b_1 = 1, b_2 = -1\right)\right\}$$

$$(5.78)$$

Further, from our discussion of the decorrelator, we know that the bit estimate of b_2 in the first stage is independent of the value of b_1. Thus, using the results from Example 5.2:

$$\Pr(\hat{b}_2 = -1|b_1 = 1, b_2 = 1) = \Pr(\hat{b}_2 = 1|b_1 = 1, b_2 = -1)$$

$$= Q\left(\sqrt{\frac{2E_b}{N_o}(1 - \rho^2)}\right)$$

$$\Pr(\hat{b}_2 = 1|b_1 = 1, b_2 = 1) = \Pr(\hat{b}_2 = -1|b_1 = 1, b_2 = -1)$$

$$= 1 - Q\left(\sqrt{\frac{2E_b}{N_o}(1 - \rho^2)}\right) \quad (5.79)$$

Now, if the estimates of b_2 are correct we have

$$\Pr(\hat{b}_1 = -1 \mid b_1 = 1, b_2 = 1, \mathrm{sgn}(y_2) = 1) = \Pr(\hat{b}_1 = -1 \mid b_1 = 1, b_2 = -1, \mathrm{sgn}(y_2) = -1)$$

$$= Q\left(\sqrt{\frac{2E_b}{N_o}}\right) \tag{5.80}$$

However, if the estimates for b_2 are not correct, the resulting impact on the error probability depends on whether or not b_2 has the same sign as b_1 or is opposite in sign. Specifically, if the bits have the same sign, (assuming ρ is positive) errors in the estimate of b_2 will actually reinforce b_1, whereas if the bits have different signs, the error in the estimate of b_2 will negate b_1:

$$\Pr\left(\hat{b}_1 = -1 \mid b_1 = 1, b_2 = 1, \mathrm{sgn}(y_2) = -1\right) = Q\left(\sqrt{\frac{2E_b}{N_o}(1 + 2\rho^2)}\right)$$

$$\Pr\left(\hat{b}_1 = -1 \mid b_1 = 1, b_2 = 1, \mathrm{sgn}(y_2) = 1\right) = Q\left(\sqrt{\frac{2E_b}{N_o}(1 - 2\rho^2)}\right) \tag{5.81}$$

Putting together (5.79)–(5.81) we have a close form expression for the probability of bit error:

$$P_b = \frac{1}{2}\left\{ Q\left(\sqrt{\frac{2E_b}{N_o}(1 + 2\rho)^2}\right) Q\left(\sqrt{\frac{2E_b}{N_o}(1 - \rho^2)}\right)\right\} + \dots$$

$$\frac{1}{2}\left\{ Q\left(\sqrt{\frac{2E_b}{N_o}}\right)\left(1 - Q\left(\sqrt{\frac{2E_b}{N_o}(1 - \rho^2)}\right)\right)\right\} + \dots$$

$$\frac{1}{2}\left\{ Q\left(\sqrt{\frac{2E_b}{N_o}}\right)\left(1 - Q\left(\sqrt{\frac{2E_b}{N_o}(1 - \rho^2)}\right)\right)\right\} + \dots$$

$$\frac{1}{2}\left\{ Q\left(\sqrt{\frac{2E_b}{N_o}(1 - 2\rho^2)}\right) Q\left(\sqrt{\frac{2E_b}{N_o}(1 - \rho^2)}\right)\right\}$$

$$= Q\left(\sqrt{\frac{2E_b}{N_o}}\right)\left(1 - Q\left(\sqrt{\frac{2E_b}{N_o}(1 - \rho^2)}\right)\right) + \dots$$

$$\frac{1}{2}Q\left(\sqrt{\frac{2E_b}{N_o}(1 - \rho^2)}\right)\left\{ Q\left(\sqrt{\frac{2E_b}{N_o}(1 - 2\rho)^2}\right) + Q\left(\sqrt{\frac{2E_b}{N_o}(1 + 2\rho)^2}\right)\right\} \tag{5.82}$$

The probability of error for two-stage receivers with either a conventional matched filter first stage or a decorrelating first stage are plotted versus positive values of ρ between 0 and 1

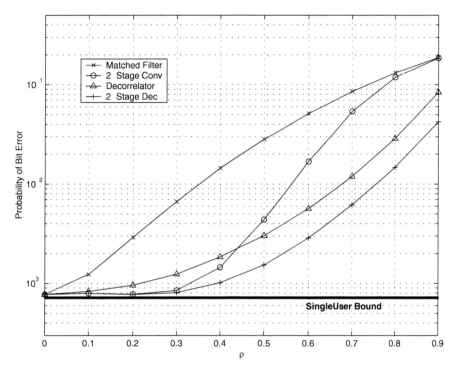

FIGURE 5.9: Bit error rate versus correlation, ρ, One and Two-Stage Detectors with Conventional Matched Filter and Decorrelator as the First Stage ($E_b/N_o = 7$dB)

in Figure 5.9. Also plotted are the BER performance of the first stages (matched filter and decorrelator). We can see that due to the improved reliability of the decisions in the first stage, the two-stage receiver with a decorrelating first stage provides improved performance over a conventional first stage receiver. Additionally, we can see that the two-stage receiver with a conventional first stage out-performs the standard decorrelator for low correlation values, but has inferior performance as the correlation grows.

As could be guessed from the preceding development, the performance of the multi-stage receiver is, in general, difficult to derive. Thus, simulations are almost exclusively used to determine performance. Additionally, the behavior as the number of stages grows is difficult to predict and doesn't always improve as the number of stages increases. In fact, the performance can degrade as the number of stages increases if the reliability of the decisions gets worse. One way to improve this is to use the concept of *partial cancellation* [112, 115]. The idea behind partial cancellation or *selective cancellation* is to attempt cancel a portion of the estimated interference, especially when the decisions are less reliable.

There are two typical techniques for implementing partial or selective cancellation. The first is to only cancel those bits which appear to be reliable. That is

$$\hat{b}_k^{(s)} = \text{sgn}\left[y_k - \sum_{j \neq k} A_j \rho_{jk} \varpi \left(z_j^{(s-1)} \right) \hat{b}_j^{(s-1)} \right] \qquad (5.83)$$

where

$$\varpi \left(z_j^{(s-1)} \right) = \begin{cases} 1 & \left| z_j^{(s-1)} \right| > \nu \\ 0 & else \end{cases} \qquad (5.84)$$

is the selective function defined for some threshold ν. We shall refer to this as *selective cancellation*.

A second approach is to partially cancel all estimates with a partial cancellation factor that increases with the cancellation stage. This reflects the fact that estimates in the early stages are less reliable than estimates in the later stages and mitigates the negative impact of canceling incorrect bit estimates. In other words, the bit estimate is formulated as

$$\hat{b}_k^{(s)} = \text{sgn}\left[y_k - \sum_{j \neq k} A_j \rho_{jk} \zeta^{(s)} \hat{b}_j^{(s-1)} \right] \qquad (5.85)$$

where $\zeta^{(s)}$ is a partial cancellation factor for stage s.

As an example, consider a system which uses random spreading codes of length $N = 100$. The simulated BER for the matched filter and up to two stages of cancellation (i.e., a three-stage receiver) are plotted in Figure 5.10 versus the number of users in the system. The user signals are assumed to be perfectly power controlled ($P_i = P \forall i$), in phase and synchronous with E_b/N_o = 8dB. We can see that the performance advantage of two stages of cancellation is substantial. For a required BER of 10^{-3}, if a matched filter receiver is used, the system can support approximately 25 simultaneous users. With two stages of full cancellation, approximately 45 users can be supported. However, by applying partial cancellation with a partial cancellation factor of 0.6 in the first stage of cancellation and 0.8 in the second stage of cancellation, the system can support approximately 60 users, a 33% improvement over standard full cancellation. Again, this benefit is derived from the fact that partial cancellation mitigates the impact of incorrect decisions which are more frequent in the early stages of cancellation [112]. In general, letting the number of stages grow does not continue to improve performance. However, the use of partial cancellation allows for more stages since it mitigates negative feedback. Another case where the performance converges as the number of stages increases is the linear cancellation case which can be thought of as partial cancellation with the partial cancellation factor equal to the matched filter magnitude. We will examine this convergence in the following example.

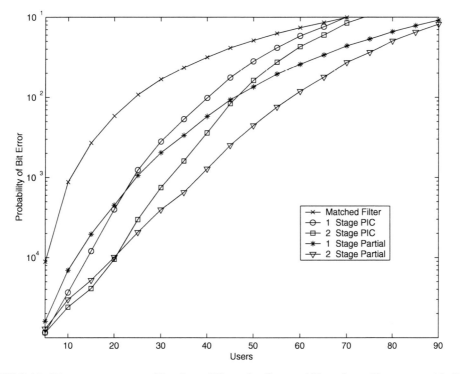

Probability of Bit Error

Users

FIGURE 5.10: Bit error rate versus Number of Users for One and Two-Stage Detectors with Conventional Full and Partial Cancellation ($E_b/N_o = 8$dB, random codes, $N = 100$)

Example 5.6. Determine the impact of letting the number of stages of a linear PIC receiver approach infinity.

Solution: For linear PIC, we can write the vector of decision statistics after one stage of cancellation as

$$
\begin{aligned}
\mathbf{y}^{(1)} &= \mathbf{RAb} + \mathbf{n} - \mathbf{Py}^{(0)} \\
&= (\mathbf{I} - \mathbf{P})\mathbf{RAb} + (\mathbf{I} - \mathbf{P})\mathbf{n}
\end{aligned}
\tag{5.86}
$$

where $\mathbf{P} = \mathbf{R} - \mathbf{I}$.   After two stages of cancellation, we have

$$
\mathbf{y}^{(2)} = (\mathbf{I} - \mathbf{P} + \mathbf{P}^2)\mathbf{RAb} + (\mathbf{I} - \mathbf{P} + \mathbf{P}^2)\mathbf{n}
\tag{5.87}
$$

Generalizing, after M stages of cancellation, the decision statistic can be written as [116]

$$
\mathbf{y}^{(M)} = \left(\sum_{s=0}^{M} (-1)^s \mathbf{P}^s \right) \mathbf{RAb} + \left(\sum_{s=0}^{M} (-1)^s \mathbf{P}^s \right) \mathbf{n}
\tag{5.88}
$$

If we let $M \to \infty$,

$$\lim_{M \to \infty} \left(\sum_{s=0}^{M} (-1)^s \mathbf{P}^s \right) = \mathrm{R}^{-1} \qquad (5.89)$$

provided that $\|\mathbf{P}\|_p < 1$ where $\|\mathbf{P}\|_p$ is the p-norm of matrix $\mathbf{P}$. Thus, as the number of stages approaches infinity, the linear PIC detector approaches the decorrelating detector. Additionally, one can view linear PIC as an implementation of Jacobi iterations for solving linear systems [116]. Interestingly, a multistage linear SIC receiver can also be shown to approach the decorrelating detector if the number of stages approaches infinity since linear SIC can be viewed as an implementation of Gauss–Seidel iterations for solving linear systems [116].

5.5   A COMPARISON OF SUB-OPTIMAL MULTIUSER RECEIVERS

In this section, we compare the BER performance of the various multiuser receiver structures. Specifically, we are interested in the ability of the receivers to mitigate multi-access interference and their ability to handle the near-far problem. We first examine the ability to mitigate multi-access interference in AWGN channels with perfect power control (i.e., all received powers being equal). Secondly, we examine near-far performance by examining the ability of each receiver structure to handle a single strong interferer. Finally, we examine realistic channel impairments such as Rayleigh fading and timing synchronization errors.

5.5.1   AWGN Channels

We first present the performance (theoretical and simulation) for AWGN channels [117]. The first set of results are capacity curves (i.e., performance versus the number of users in the system) for $E_b/N_o = 8$dB, $N = 31$, and perfect power control. The simulation results are plotted along with the theoretical curves in Figure 5.11. The parallel scheme uses two stages of cancellation ($S = 3$) and a partial cancellation factor of 0.5 in stage 2 [112]. The simulation results and the theoretical results agree and show similar trends.

For the perfect power control case, we find that the decorrelator, MMSE, parallel interference canceller, and decorrelating decision-feedback (DF) detectors all provide similar performance although the latter two are slightly better. The successive interference canceller performs significantly worse than the other three receivers due to the lack of variance in the received powers. In fact, the performance is only insignificantly better than the conventional receiver. One important aspect of this figure is that it plots BER performance averaged over all users. For most of the detectors, the performance of any specific user is equal to the average performance. However, this is not true for the successive interference cancellation receiver. The average performance in this case is dominated by the performance of the first detected user

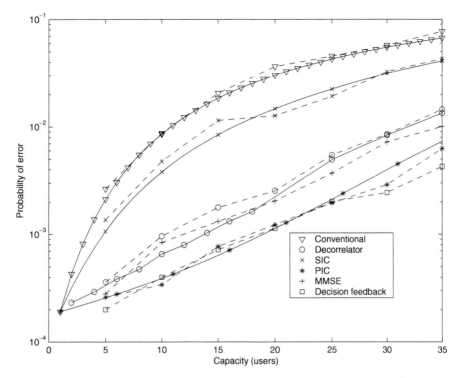

FIGURE 5.11: Bit error rate versus number of users for perfect power control ($E_b/N_o = 8$dB and processing gain $= 31$; *solid lines* represent analytical results and *dashed lines* represent simulated results)

which is equivalent to the performance of the conventional matched filter receiver. The performance versus E_b/N_o is given in Figure 5.12 for $K = 10$, $N = 31$, and perfect power control. Again, we find significant improvement for the decorrelator, the parallel interference canceller, the MMSE, and decorrelating DF receivers, with each providing gains over the matched filter of over an order of magnitude at $E_b/N_o = 10$dB, while the successive canceller provides a small improvement.

5.5.2   Near-Far Performance

As mentioned earlier, one of the drawbacks of the conventional receiver is that it is subject to the near-far problem. One means of characterizing the robustness of a multiuser detector to the near-far problem is the measure developed by Verdú [119] termed *near-far resistance*. Near-far resistance is based on the concept of *effective energy*. Effective energy is the energy required by the matched filter in the presence of only AWGN to obtain the same BER as a particular multiuser detector operating in the presence of AWGN *and* multi-access interference. Formally, let us fix the multi-access interference experienced by a particular signal and define the BER of

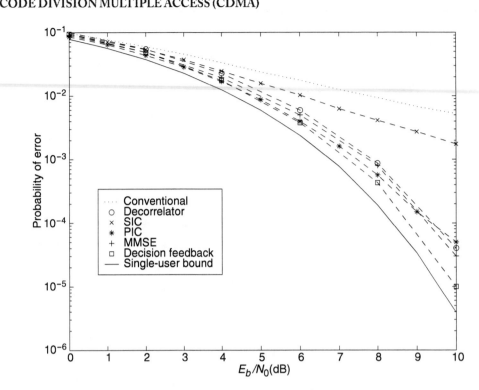

FIGURE 5.12: BER versus E_b/N_o with perfect power control (10 users and processing gain $= 31$)

a multi-user detector as a function of noise power as $P_e(\sigma)$. The energy required by a matched filter to achieve the same BER in the absence of multi-access interference with the same noise power is the effective energy $e_k(\sigma)$. That is

$$P_e(\sigma) = Q\left(\sqrt{\frac{e_k(\sigma)}{\sigma^2}}\right) \qquad (5.90)$$

Clearly, $e_k(\sigma) \le A_k^2$ or in other words, the effective energy is upper-bounded by the actual energy. Another way of viewing this is that the matched filter in the absence of interference cannot require *more* energy to achieve the same BER as any multi-user detector in the presence of multi-access interference. If a detector achieves equality in terms of the effective energy, we can interpret this as perfectly eliminating the multi-access interference. Multiuser efficiency [55, 119] is the ratio of effective energy to actual energy:

$$\eta_k(\sigma) = \frac{e_k(\sigma)}{A_k^2} \qquad (5.91)$$

Asymptotic multiuser efficiency is the limit of multiuser efficiency as the noise level goes to zero. That is

$$\eta_k = \lim_{\sigma \to 0} \frac{e_k(\sigma)}{A_k^2} \qquad (5.92)$$

This provides the rate at which the BER of a specific detector in the presence of fixed multi-access interference goes to zero as the noise power goes to zero. Finally, near-far resistance can be defined as the worst case asymptotic multiuser efficiency over all possible interference energies:

$$\bar{\eta}_k = \inf_{A_j, j \neq k} \frac{e_k(\sigma)}{A_k^2} \qquad (5.93)$$

A second, less formal, approach to examine near-far performance is to plot the performance of a receiver structure in the presence of two interferers, one with equal power to the desired user and one with a power that grows to an extremely large level. Figure 5.13 presents the simulated performance of various detectors as the power of one interferer grows from 10dB below the

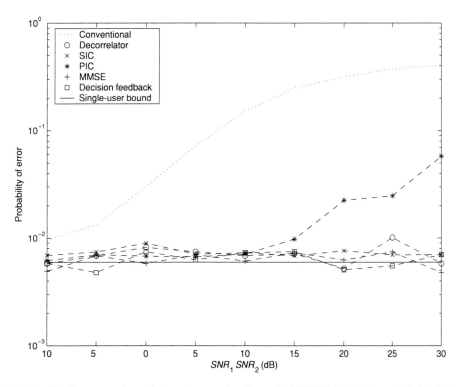

FIGURE 5.13: Performance degradation in near-far channels (AWGN, $E_b/N_o = 5$dB for desired user and spreading gain $= 31$)

desired user to 30dB above the desired user for $N = 31$ and $E_b/N_o = 5$dB. As expected from our discussion in Chapter 2, the conventional receiver degrades quickly in the presence of strong interference. The successive interference cancellation receiver benefits from diverse powers and is found to be robust to the strong interferer. This is intuitive, since the SIC receiver detects and cancels signals in decreasing order of received signal strength. As the stronger signal (which is detected first) continues to grow in received power, it is detected and cancelled more effectively thus improving the performance of the successively detected signals.

In section 5.3.1, we saw that the performance of the decorrelator is independent of the received signal strength of the interfering signals as shown in equation (5.24). This is also seen in Figure 5.13. Additionally, the figure shows that the MMSE and decorrelating DF receivers show similar robustness to the near-far problem. The parallel cancellation receiver is less robust and shows slow degradation for high interference power. The parallel canceller suffers because cancellation of the weak signal is inaccurate in the first stage of cancellation due to the dominating interference. This poor cancellation serves to degrade the estimate of the strong signal in the succeeding stage. Consequently, when the strong signal is cancelled from the weak signal in second stage of cancellation, it is done inaccurately, degrading the detection of the signal. This continues from stage to stage with slight improvement each time. However, as we found in Example 5.6, for a linear PIC detector as $s \to \infty$, the parallel scheme approaches the decorrelator will thus be near-far resistant. However, due to inaccurate channel estimation, the parallel cancellation approach is in general not near-far resistant [118].

5.5.3   Rayleigh Fading

To examine the performance of the receiver structures in a more realistic channel (rather than simple AWGN), simulations were performed for each detector in fading channels. While typically the assumption of Rayleigh fading is somewhat pessimistic for wideband channels,[1] a Rayleigh fading model are more easily compared with previous results. The performance results for each of the receiver structures in flat Rayleigh fading are presented in Figure 5.14. It is further assumed that the fading is slow (a coherence time of several bit intervals) and that the phase can be tracked with sufficient accuracy. Again, we find significant improvement over the conventional receiver with each of the receivers providing nearly equivalent performance. As in the AWGN case, the performance is extremely close to the single-user bound. The cancellation techniques do seem to have a slight disadvantage, which is likely due to the need for channel

---

[1]As the bandwidth of a system increases, the number of resolvable multipath components increases. It has been noted that as more paths become resolvable, these paths are no longer Rayleigh distributed [7, 120]. Rather, the Rayleigh fading effect is mitigated, resulting in a reduced signal strength variance.

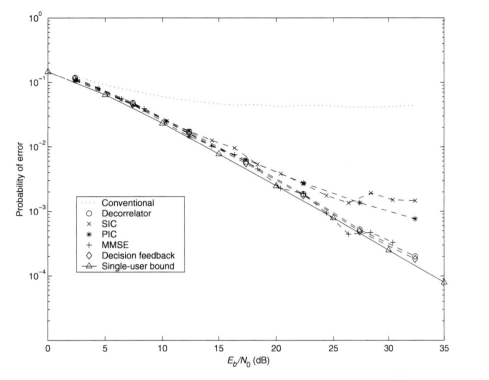

FIGURE 5.14: BER versus E_b/N_o for flat Rayleigh fading (10 users, spreading gain $= 31$, $\acute{O} = 0.93$, coherence time $= 50$ bit intervals)

gain estimates. It should be noted that SIC receiver performs well in Rayleigh fading, unlike the AWGN case. This is due to the natural power variation caused by independent fading.

5.5.4 Timing Estimation Errors

We have assumed up to this point that the delays of each path for each user are known exactly. In a realistic system, some delay estimation error will be present. Thus, it is useful to observe the effect of delay estimation errors on the performance of each of the multiuser detectors [121]. The estimation error is assumed to be Gaussian with some standard deviation in chips (or fraction of a chip). The effects of delay estimation error on the performance of each of the receivers studied for $E_b/N_o = 8$dB, $N = 31$, and $K = 20$ in a perfect power control AWGN channel are shown in Figure 5.15. As expected, there comes a point where the attempted removal of interference becomes no longer useful and even harmful. We can also see that the performance degrades very rapidly for delay estimation errors. For an error standard deviation of only one-tenth of a chip, performance is degraded by more than an order of magnitude. These results show that timing errors are more critical for multiuser receivers than they are for matched filters since

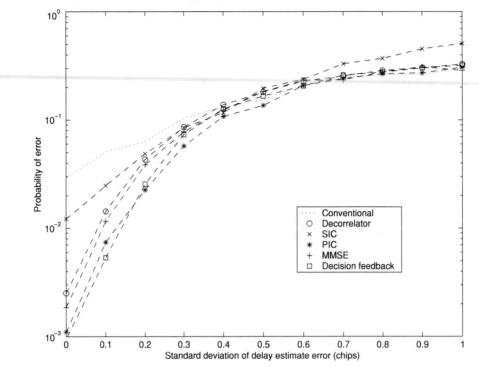

FIGURE 5.15: Effect of timing errors (delay estimate errors) on system performance in an AWGN channel with perfect power control ($E_b/N_o = 8$dB, spreading gain $= 31$, $K = 20$)

we not only lose correlation energy, but also perform increasingly inaccurate cancellation. The combination of these two effects makes timing more critical. Thus, timing will be a significant design issue for multiuser implementation. Figure 5.16 shows the effect of timing errors in near-far channels and demonstrates that, for even a small amount of timing error, none of the receivers can maintain near-far resistance. Therefore, for a realistic system, loose power control will still be necessary.

5.6   APPLICATION EXAMPLE: IS-95

We would like to now examine factors that impact the implementation of multiuser detection in real-world systems. Specifically, we will examine the cellular CDMA standard termed IS-95 [122]. The conventional receiver for IS-95–based CDMA systems [122] is a four-finger Rake receiver using filters matched to a single user's spreading code on each finger, equal gain combining, and square-law detection. Additionally, the conventional IS-95 base station typically uses two-antenna for receive diversity. It has been argued that a design philosophy that seeks to randomize the interference as much as possible is the best approach and that any structure added

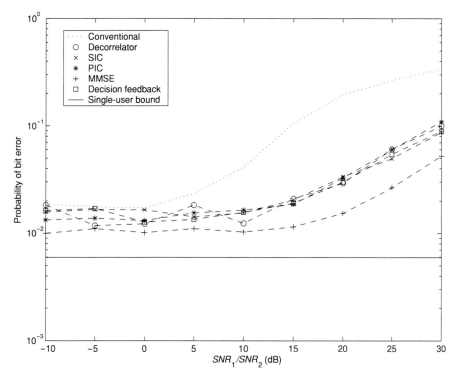

FIGURE 5.16: Effect of timing errors (delay estimate errors) on system performance in near-far channels (AWGN, $E_b/N_o = 5$dB, spreading gain $= 31$, $K = 10$)

to allow multiuser techniques will degrade overall system performance [123]. Following that philosophy, the IS-95 uplink uses long pseudo-random spreading sequences. As we discussed in Section 5.3, linear receivers use knowledge of the spreading codes[2] of all users to create a linear transformation to project each signal into an orthogonal subspace. Specifically, the decorrelator removes the effects of interference by projecting the signal of interest onto a subspace that is orthogonal to the entire interference subspace. Since this subspace is not in the same direction as the desired signal, the receiver suffers a loss of desired signal energy, and thus the performance versus thermal noise degrades. The MMSE receiver projects the desired signal in the direction that minimizes the combined effect of interference and thermal noise. The additional benefit of this receiver structure is that it can be implemented adaptively and blindly. Unfortunately, linear multiuser receiver structures are incompatible with IS-95 for several reasons. Since the reverse link uses 64-ary orthogonal modulation [122], subspace methods are extremely limited [55]. Specifically, since each received signal must be projected into 64 orthogonal directions

---

[2] Actually, the MMSE receiver can be implemented without knowledge of the interferers' spreading sequences.

to make symbol decisions and the received signal can have at most $64 \times 4 = 256$ dimensions[3], linear subspace methods are severely constrained.

Even if the reverse link modulation in IS-95 were changed to a linear scheme with a smaller number of dimensions (e.g.,BPSK as in the *cdma2000* standard), adaptive cancellation (a significant advantage of linear detection) could not be used because of the use of long spreading codes, which changes the interference subspace every symbol. Thus, the linear filter would need to be recomputed every symbol interval (a severe computational burden), and adaptive techniques would be impossible. Resolvable multipath further limits this technique since each additional multipath will occupy a signal dimension. For example, if there are four resolvable multipath components, the number of users that can be projected into orthogonal dimensions is decreased by a factor of four.

The decorrelating decision feedback detector suffers from the same limitations as the linear detectors. However, PIC and SIC receivers (or multistage implementations of them) are compatible with any modulation scheme since they rely on regeneration and cancellation of the interference. Thus, these receiver structures are applicable for IS-95. However, these structures also encounter a challenge when implemented in IS-95 [123, 124]. Cancellation techniques encounter difficulty because cancellation operates on coded symbols and the coded symbol SINR is often too low to make reliable decisions. More reliable coded symbol estimates could be obtained at the output of a decoder, but this introduces substantial memory requirements and a significant delay, which may be unacceptable for two-way voice communications. In addition, the low SNR combined with the fluctuating level of the received signal power caused by the mobile environment make reliable channel estimation difficult, which is critical in cancellation approaches. Furthermore, cancellation can be applied only to interference that is known. Out-of-cell interference (OCI) is not detected by the base station of interest and thus cannot be cancelled. In addition, OCI is likely to be too weak for reliable cancellation even if information were available.

So, what can be said then about the usefulness of interference cancellation? First, while coding certainly drives down the coded SNR, it cannot drive it down arbitrarily far. Most powerful coding techniques cannot provide gains at input error rates higher than about 10–20%. Even at a coded symbol error rate of 10%, applying brute force cancellations reduces interference by approximately 80%.[4] While imperfect channel estimation further limits the improvement, a soft cancellation approach can minimize the effects of symbol decision errors by weighting incorrect decisions according to their reliability [125, 126]. Furthermore, our experience shows that the typical wireless channel remains relatively constant over an IS-95 power control group

---

[3]Since the spreading codes have four chips per Walsh chip or 256 chips per symbol, the dimensionality of the received signal is 256.

[4]This assumes that the 10% error rate increases the interference.

(1.25ms). This corresponds to 6 Walsh symbols in IS-95, allowing multiple channel observations per estimate and a corresponding 7dB improvement in channel estimation (assuming correct symbol decisions). Most importantly, while the reliability of the coded data and the channel estimates may be relatively poor in the first stage of cancellation (in a multistage approach), intelligent cancellation improves the reliability in the following stages.

5.6.1 Parallel Interference Cancellation

The previous discussion motivates the examination of interference cancellation techniques in IS-95 systems. While SIC is technically applicable, we focus on PIC in this discussion for the following reasons. First, PIC lends itself naturally to parallel implementation. SIC, on the other hand, must be done sequentially, implying a much more difficult implementation. Additionally, PIC lends itself more naturally to a multistage approach, which will prove to be useful when the initial estimates are not very reliable. Other issues, such as power control, also have an effect on the cancellation technique. To obtain equal BERs, SIC requires a geometric distribution as demonstrated in Example 5.5 while PIC requires equal powers.

We describe PIC by presenting the complex baseband representation of the received signal at the ith antenna as

$$r_i(t) = \sum_{K=1}^{K} \sum_{i=1}^{L_k} \gamma_{k,i,l}(t) w_k(t - \tau_{k,l}) a_k(t - \tau_{k,l}) + n_i(t) \qquad (5.94)$$

where $\gamma_{k,i,l}(t) = \alpha_{k,i,l}(t) e^{j\theta_{k,i,l}}(t)$ is the multiplicative distortion (both amplitude and phase) seen by the lth resolvable path of the kth user's signal at the ith antenna, $w_k(t)$ is the Walsh function of the kth user that carries the data, $a_k(t)$ is the complex spreading sequence of the kth user representing both the long and sort codes, $n_i(t)$ is complex Gaussian noise that has variance σ_n^2 in in-phase and quadrature and is assumed to be spatially and temporally white, $\tau_{k,l}$ is delay seen by the lth path of the kth user that is assumed to be large compared to the propagation time across the array, and L_k represents the number of resolvable paths in the kth user's received signal.

In the conventional receiver, detection of the kth user's signal is accomplished by despreading the received signal by the complex conjugate of the kth user's spreading code and subsequently taking the Walsh transform ($W\{x\}$) of each diversity path, i.e., over each antenna and resolvable multipath. Each transform will result in a length 64 vector. That is,

$$\overline{Z}_{k,l,i,n} = W\{r_i(t) a_k^*(t - \tau_{k,j}); n, \tau_{k,j}\}$$

$$= \int_{(n-1)T_i + \tau_{kj}}^{nT_i + \tau_{k,j}} r_i(t) a_k^*(t - \tau_{k,l}) * (t - \tau_{k,j}) dt \qquad (5.95)$$

where $Z_{k,l,i,n}$ is the vector of Walsh transform outputs for the lth path of the kth user received on the ith antenna during the nth symbol interval, $*(t)$ is the vector of orthogonal Walsh functions, and $(\cdot)^*$ represents the complex conjugate. We now drop the dependence upon n for notational convenience and express the decision statistic as the non-coherent vector sum over diversity (antenna and multipath) vectors, i.e.,

$$\overline{Z}_k = \sum_l \sum_i \left| \overline{Z}_{k,i,l} \right|^2 \qquad (5.96)$$

The estimated Walsh symbol is then chosen as the one that corresponds to the index of the largest value of Z_k.

$$\hat{w}_k(t) = \sum_{n=-\infty}^{\infty} w^{(m[n])}(t - nT_i) \qquad (5.97)$$

where $m[n]$ corresponds to the index that contains the largest value during each symbol interval. In a conventional system, the Walsh outputs are then used to create bit metrics that are fed to the soft-decision Viterbi decoder. As mentioned, interference cancellation occurs prior to decoding. Thus, the decisions made by the matched filter can be used along with channel estimates to recreate and cancel interference to each user. The new received signal on the ith antenna for the lth path of the kth user can be represented by

$$r_i^{(k,l)}(t) = r_i(t) - \sum_j \sum_{\substack{m \neq l \\ \text{if} \\ j = k}} \hat{\gamma}_{j,i,m}\hat{w}_j(t - \hat{\tau}_{j,m})a_j(t - \hat{\tau}_j, m) \qquad (5.98)$$

Note that while we represent a different new received signal for each path of each user for conceptual clarity, in practice we will work with a single residual signal [126]. Once interference cancellation has been completed for each user, the new received signals $r_i^{(k,j)}(t)$ are used in detection as before. That is,

$$\overline{Z}_{k,l,i}^{(1)} = W \left\{ r_i^{(k,l)}(t)a_k^*(t - \tau_{k,l}); \tau_{k,l} \right\} \qquad (5.99)$$

where we use the superscript (1) to denote one stage of cancellation. This new estimate can then be used along with improved channel estimates to re-estimate and cancel the interference, allowing another stage of estimation. The number of useful stages is a function of loading. For a lightly loaded system, one stage of cancellation may obtain 99% of the achievable gain, and heavily loaded systems may require three or four stages of cancellation. We shall add the superscript (s) to represent the number of stages. The signal used for detection of the lth path of the kth user on the ith antenna after s stages cancellation will thus be represented by $r_i^{k,l,s}$.

5.6.2 Performance in an Additive White Gaussian Noise Channel

Our initial investigation of PIC for IS-95 focuses on the simplest case: an AWGN channel. Figure 5.17 shows the simulated performance of PIC in an AWGN channel as the number of active users grows. BER is plotted against system loading. Note that voice activity, coding, power control, and OCI are not considered. The channel is estimated using a six-symbol average of Walsh outputs. From Figure 5.17, we see that for a target uncoded BER of 1%, nine stages of interference cancellation can increase cell capacity nearly 5 times. A single stage of cancellation gets nearly half of that improvement while four stages of cancellation obtain nearly all of it.

Channel estimation is important for interference cancellation for obvious reasons. The estimation of the channel can be approached several ways. From (5.94) and (5.95), we can express the wth element of the Walsh output vector as

$$Z_{k,l,i,w} = \begin{cases} T_s \Gamma_{k,l,i} + \sum_j \sum_{\substack{m \neq l \\ \text{if} \\ j=k}} \Gamma_{j,m,i} I_{j,k,i,m} + \mathcal{N}_{k,j,i} & w = w_{max} \\ \sum_j \sum_{\substack{m \neq l \\ \text{if} \\ j=k}} \Gamma_{j,m,i} I_{j,k,l,m} + \mathcal{N}_{k,l,i} & w \neq w_{max} \end{cases} \quad (5.100)$$

where $\Gamma_{k,l,i}$ is the channel of the lth path of the kth user received on the ith antenna after integration, $I_{j,k,l,m}$ is the correlation between the lth path of the kth user and the mth path of

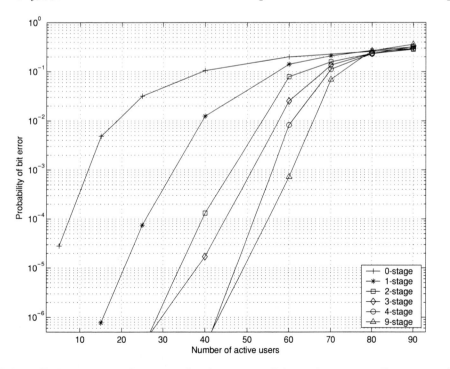

FIGURE 5.17: Bit error rate performance of multistage parallel interference cancellation in an AWGN channel versus system loading ($E_b/N_o = 8$dB, no coding)

the jth user (i.e., interference), N is the post-correlation AWGN term, and w_{max} is the index of Walsh vector value with the largest Rake combined energy. If we can model the interference term as AWGN [121, 127] (i.e., a zero-mean complex Gaussian random variable), then the Walsh output with the transmitted symbol can also be used as an estimate of the channel. Three things affect this estimate: the interference and noise terms; the channel's variance over the symbol interval; and the choice of the correct symbol to obtain the estimate. The last effect is unavoidable since the modulation scheme is non-linear. In 64-ary orthogonal modulation, it is impossible to remove the effect of the modulation without a training sequence or pilot symbols. Thus, a correct decision is necessary to obtain a proper channel estimate. One method of mitigating symbol decision errors is to average over multiple symbols. While a single symbol error will certainly degrade a channel estimate based on a multiple-symbol observation interval, it will not make it unusable. However, the number of Walsh symbols must not exceed a substantial fraction of the channel coherence time.

5.6.3   Multipath Fading and Rake Reception

Multipath fading will affect the performance of PIC in a number of ways. First of all, fading makes channel and symbol estimation more difficult, presenting several additional challenges to the design of a PIC receiver. As mentioned previously, the coherence time of the channel must be considered. Since cancellation must occur on individual Rake fingers, the effect of deep instantaneous drops in finger energy must also be considered. Symbol decisions are made after Rake combining, which improves the reliability of symbol estimates. Figure 5.18 plots the simulated BER performance of PIC and the conventional receiver in two-ray Rayleigh fading versus normalized system loading. The ratio of total combined bit energy to thermal noise spectral density $\overline{E_b/N_o}$ is 15dB. Two receive antennas, spatially separated by ten carrier wavelengths, are assumed. The resolvable signal components for each user are separated by 5µs, with the second arriving component 6dB lower than the first. We can see from Figure 5.18 that not only does PIC perform well in fading, but the relative gains in terms of capacity at 1% BER are even greater than those in AWGN. Thus, fading does not necessarily reduce the relative capacity gains achievable despite the channel impairments.

5.6.4   Voice Activity, Power Control, and Coding

As discussed in Chapter 3, one of the advantages of CDMA systems, and IS-95 in particular, is that voice activity is exploited to enhance capacity. During a typical conversation, a speaker is talking about 3/8 of the time [41]. Voice codecs required by IS-95 allow this fact to be exploited by reducing average mobile station transmit power by as much as 9dB when a user is not talking. The net effect is that overall interference power is reduced by about 50%. This is achieved in

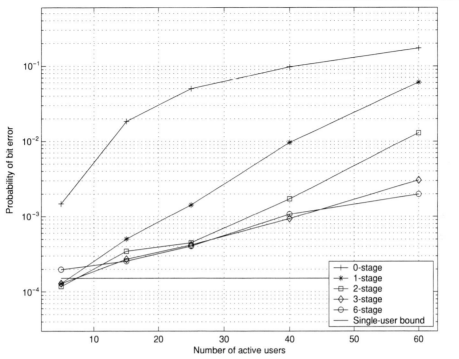

FIGURE 5.18: BER performance of multistage interference cancellation versus system loading in a two-ray Rayleigh fading channel with two-antenna diversity ($E_b/N_o = 15$dB, no coding)

IS-95 by using four different transmission rates. Each 20ms voice frame of the IS-95 reverse link is composed of sixteen (1.25ms) power control groups (PCGs) (96 total Walsh symbols). During full-rate transmission, all sixteen PCGs are transmitted. During 1/2 rate, 1/4 rate, and 1/8 rate, however, eight, four, and two PCGs are transmitted, respectively. Since the rate is unknown to the base station prior to Viterbi decoding and since PIC operates on coded symbols, the PIC must be designed to account for this effect. Canceling estimated interference of one user during a PCG that was not transmitted, for example, would cause interference to be added to rather than subtracted from the combined received signal. Cancellation, therefore, is performed on a PCG-by-PCG basis. Before performing cancellation, we must first determine whether or not each user's signal is present during a given PCG by comparing the maximum average Walsh energy over a PCG to a predetermined threshold. If the threshold is exceeded, we conclude that the user was active during the PCG in question and cancellation is performed. However, the final decision on voice rate is still not made until after decoding.

Power control and forward error control (FEC) coding are essential parts of IS-95. It is, therefore, also critical to consider these when applying interference cancellation. Power control and FEC are considered together since they are tied together in IS-95. Power control is partially based on frame errors (through the outer loop), which are determined at the output of the Viterbi decoder, operating on each 20ms frame. The two issues are important to consider in interference cancellation since they both have the tendency to drive down the input SINR, which makes both channel estimation and symbol estimation more difficult. Estimation errors in turn degrade cancellation performance. However, as mentioned earlier, partial cancellation can be performed in the early stages, if necessary, to reduce the effects of symbol errors. In addition, using a six-symbol observation for the channel estimate improves the SNR of the channel estimation by 7dB.

Because power control is partially based on FER, PIC can be implemented without affecting the power control algorithm. Cancellation will improve the SINR at the input of the Viterbi decoder for a given received SINR. This allows a lower SINR at the input of the receiver for a target FER. A lower allowable received SINR translates into a larger allowable user population, i.e., larger capacity. To determine the increase in capacity in the presence of power control, we define the capacity as the point at which power control can no longer maintain the target FER. Since power control drives the system to a target FER (assumed to be 1%), the FER performance of the conventional receiver and PIC will be the same for low system loading. As the loading increases, however, at some point the conventional receiver will be unable to maintain the target FER for all users. When this occurs, the system is unstable and assumed to be loaded beyond its capacity. If the loading level at which the PIC receiver breaks down is higher than that of the conventional receiver, PIC is said to provide a capacity increase.

5.6.5   Out-of-Cell Interference

To this point, we have not specifically addressed the effect of OCI. Often OCI is modeled by assuming a sufficiently high thermal noise level so as to include its effect. This is inadequate for a couple of reasons. First, as system loading increases, the OCI should increase proportionally. Second, if interference cancellation reduces the transmit power at the mobile, the interference level seen in surrounding cells will also reduce. To accommodate these problems, we model OCI as AWGN that has a power level proportional to the total in-cell interference. We represent this ratio by η. It is typically reported that OCI is approximately 55% of intracell interference [40]. Thus, we model OCI as AWGN that is $\eta = 0.55$ times the total received in-cell interference. This accommodates the fact that OCI should increase as the cell loading increases and should decrease as the average transmit power per mobile decreases. By modeling OCI as AWGN, we reflect the fact that we do not have information concerning out-of-cell users (i.e., we cannot cancel OCI) and that the OCI is composed of a large number of low power signals. Using the

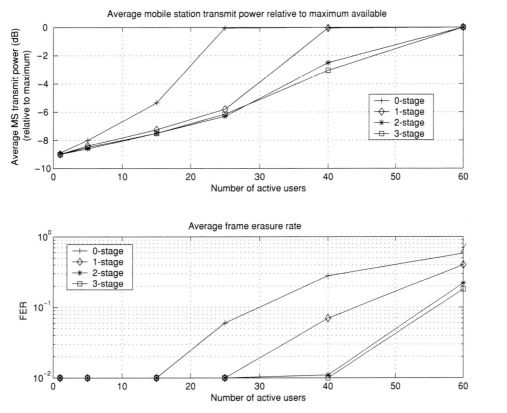

FIGURE 5.19: Performance of multistage interference cancellation in IS-95-like system ($E_b/N_o = $ 15dB, $K = 9$, $r = 1/3$ convolutional coding, closed-loop power control; OCI factor, $C = 0.55$; average voice activity $= 50\%$; maximum frequency offset is 300Hz; normalization relative to conventional capacity)

well-known Gaussian approximation [127], the OCI variance is determined to be

$$\sigma_I^2 = 0.55 \times \frac{\sum_k P_k}{3N} \qquad (5.101)$$

where P_k is the average power received from the kth user and N is the number of chips per Walsh symbol, which is 256 in IS-95.

Figure 5.19 plots the simulated results for FER and average mobile station (MS) transmit power versus system loading in a 150-Hz Rayleigh fading channel with all users exhibiting approximately 50% voice activity. The simulation assumes a two-ray Rayleigh fading channel with two receive antennas, an average combined E_b/N_o of 15dB, and OCI modeled as in (5.101). Note that a frequency offset is also included in this simulation. A random frequency offset (due to imperfect carrier demodulation) is applied to each user, where the offset is assumed to be a

Gaussian random variable with a standard deviation of 150Hz. We see that an approximately 2.5 times capacity increase is possible on the uplink, or a 2- to 5-dB reduction in MS transmit power is possible. Even when considering the effects of Rayleigh fading, frequency offset, voice activity, channel coding, power control, and intercell interference, we find that the PIC receiver provides significant benefits in system performance.

5.7   SUMMARY

In this chapter, we have described joint detection techniques that are particularly applicable to the uplink of CDMA systems. The optimal joint detection technique (also known as optimal multiuser detection), while providing substantial performance improvement, has a complexity that is exponential in the number of signals being detected. Thus, sub-optimal approaches are of interest and are typically divided into linear and non-linear techniques. Both types were thoroughly described in this chapter. Additionally, we investigated the application of non-linear multiuser detection to a common cellular CDMA standard (IS-95). While there are several complicating factors that must be considered in real-world systems, it was shown that multiuser detection can still provide substantial capacity improvement on the system uplink. However, to provide actual capacity gains, the uplink improvements must be matched with corresponding downlink improvements.

Bibliography

[1]   M. K. Simon, J. K. Omura, R. A. Scholtz, and B. K. Levitt, *Spread Spectrum Communications Handbook*, Electronic Ed. New York: McGraw-Hill, 2002.

[2]   D. Agrawal and Q.-A. Zeng, *Introduction to Wireless and Mobile Systems*. Pacific Grove, CA: Brooks/Cole, 2003.

[3]   K. I. Kim, *Handbook of CDMA System Design, Engineering, and Optimization*. Englewood Cliffs, NJ: Prentice Hall, 2000.

[4]   D. Gross and C. Harris, *Fundamentals of Queueing Theory*. New York: Wiley, 1985.

[5]   V. H. MacDonald, "The cellular concept," *Bell Systems Technical Journal*, vol. 58, pp. 15–43, January 1979.

[6]   W. C. Y. Lee, *Mobile Cellular Telecommunications Systems*. New York: McGraw-Hill, 1989.

[7]   T. S. Rappaport, *Wireless Communications: Principles and Practice*, 2nd ed. Upper Saddle River, NJ: Prentice Hall, 2002.

[8]   N. Abramson, "The ALOHA system—another alternative for computer communications," *Proceedings of the Fall Joint Computer Conference*, vol. 37, pp. 281–285, 1970.

[9]   N. Abrahmson, "Development of the ALOHANET," *IEEE Transactions on Information Theory*, vol. 31, pp. 119–123, March 1985.

[10] P. Karn, "MACA: A new channel access method for packet radio," in *Proceedings of the Ninth ARRL/CRRL Amateur Radio Computer Networking Conference*, pp. 134–140, September 1990.

[11] V. Bhargavan, "MACAW: A media access protocol for wireless LAN," in *Proceedings of the Association for Computing Machinery Special Interest Group on Data Communications*, pp. 215–225, August 1994.

[12] F. Talucci and M. Gerla, "MACA-BI (MACA By Invitation). A wireless MAC protocol for high speed ad hoc networking," in *proceedings of the Sixth IEEE International Conference on Universal Personal Communications*, pp. 913–917, 1997.

[13] F. A. Tobagi and L. Kleinrock, "Packet switching in radio channels: Part II—The hidden terminal problem in carrier sense multiple-access and the busy-tone solution," *IEEE Transactions on Communications*, vol. 23, pp. 1417–1433, December 1975.

[14]　F. A. Tobagi and L. Kleinrock, "Packet switching in radio channels: Part III—Polling and (dynamic) split-channel reservation multiple access," *IEEE Transactions on Communications*, vol. 24, pp. 823–844, December 1976.

[15]　C. Wu and V. O. K. Li, "Receiver-initiated busy-tone multiple access in packet radio networks," in *Proceedings of the Association for Computing Machinery Special Interest Group on Data Communications*, pp. 336–342, 1987.

[16]　C. L. Fullmer and J. J. Garcia-Luna-Aceves, "Solutions to hidden terminal problems in wireless networks," in *Proceedings of the Association for Computing Machinery Special Interest Group on Data Communications*, pp. 39–49, 1997.

[17]　Z. J. Haas and J. Deng, "Dual busy tone multiple access (DBTMA)—a multiple access control scheme for ad hoc networks," *IEEE Transactions on Communications*, vol. 50, pp. 975–985, June 2002.

[18]　"Wireless LAN MAC and physical layer specifications," Tech. Rep., IEEE 802.11, June 1997.

[19]　R. L. Peterson, R. E. Ziemer, and D. E. Borth, *Introduction to Spread Spectrum Communications*. Englewood Cliffs, NJ: Prentice Hall, 1995.

[20]　D. Torrieri, *Principles of Spread-Spectrum Communication Systems*. New York: Springer, 2005.

[21]　A. Papoulis, *Probability, Random Variables, and Stochastic Processes*, 3rd ed. New York: McGraw-Hill, 1991.

[22]　J. G. Proakis, *Digital Communications*, 3rd ed. New York: McGraw-Hill, 1995.

[23]　D. V. Sarwate and M. B. Pursley, "Crosscorrelation properties of pseudo-random and related sequences," *Proceedings of the IEEE*, vol. 68, pp. 593–619, May 1980.

[24]　R. Gold, "Optimum binary sequences for spread spectrum multiplexing," *IEEE Transactions on Information Theory*, vol. 13, pp. 619–621, October 1967.

[25]　E. H. Dinan and B. Jabbari, "Spreading codes for direct sequence CDMA and wideband CDMA cellular networks," *IEEE Communications Magazine*, vol. 36, pp. 48–54, September 1998.

[26]　T. Kasami, "Weight distribution formula for some class of cyclic codes," University of Illinois – Urbana, Tech. Rep. R-285, April 1966.

[27]　L. R. Welch, "Lower bounds on the maximum cross-correlation of signals," *IEEE Transactions on Information Theory*, vol. 20, pp. 397–399, May 1974.

[28]　M. Z. Win and R. A. Scholtz, "Impulse radio: How it works," *IEEE Communications Letters*, vol. 2, no. 2, pp. 36–38, 1998.

[29]　D. Hibbard, "On the impact of bandwidth on the performance of indoor wireless systems," M.S. thesis, Virginia Tech, Blacksburg, VA, May 2004.

[30] J. S. Lehnert, "An efficient technique for evaluating direct-sequence spread-spectrum communications," *IEEE Transactions on Communications*, vol. 37, pp. 851–858, August 1989.

[31] J. Lehnert and M. B. Pursley, "Error probabilities for binary direct-sequence spread-spectrum communications with random signature sequences," *IEEE Transactions on Communications*, vol. 35, pp. 87–98, January 1987.

[32] R. K. Morrow and J. S. Lehnert, "Bit-to-bit error dependence in slotted DS/SSMA packet systems with random signature sequences," *IEEE Transactions on Communications*, vol. 37, pp. 1052–1061, October 1989.

[33] J. M. Holtzman, "A simple, accurate method to calculate spread-spectrum multiple access error probabilities," *IEEE Transactions on Communications*, vol. 40, pp. 461–464, March 1992.

[34] J. C. Liberti and T. S. Rappaport, "Analytical results for capacity improvements in CDMA," *IEEE Transactions on Vehicular Technology*, vol. 43, pp. 680–690, August 1994.

[35] E. A. Geraniotis and M. B. Pursley, "Error probabilities for slow-frequency-hopped spread-spectrum multiple-access communications over fading channels," *IEEE Transactions on Communications*, vol. 30, pp. 996–1009, May 1982.

[36] K. Cheun and W. E. Stark, "Probability of error in frequency-hop spread-spectrum multiple-access communication systems with noncoherent reception," *IEEE Transactions on Communications*, vol. 39, pp. 1400–1410, September 1991.

[37] Y. Yeh and S. C. Schwartz, "Outage probability in mobile telephony due to multiple log-normal interferers," *IEEE Transactions on Communications*, vol. 32, pp. 380–388, April 1984.

[38] M. Romeo, V. Da Costa, and R. Bardou, "Broad distribution effects in sums of log-normal random variables," *European Physical Journal B*, vol. 32, pp. 513–525, 2003.

[39] M. Alouini and M. K. Simon, "Dual diversity over correlated log-normal fading channels," *IEEE Transactions on Communications*, vol. 50, pp. 1946–1959, December 2002.

[40] A. J. Viterbi, "The orthogonal-random waveform dichotomy for digital mobile personal communication," *IEEE Personal Communications Magazine*, vol. 1, pp. 18–24, First Quarter 1994.

[41] P. T. Brady, "A statistical analysis of on–off patterns in 16 voice conversations," *Bell Systems Technical Journal*, vol. 47, no. 1, pp. 73–91, 1968.

[42] K. S. Gilhousen *et al.*, "On the capacity of a CDMA system," *IEEE Transactions on Vehicular Technology*, vol. 40, pp. 303–312, May 1991.

[43] M. B. Pursley, "The role of spread spectrum in packet radio networks," *Proceedings of the IEEE*, vol. 75, pp. 116–134, January 1987.

[44] E. S. Sousa and J. A. Silvester, "Spreading code protocols for distributed spread-spectrum packet radio networks," *IEEE Transactions on Communications*, vol. 36, pp. 272–281, March 1988.

[45] D. Raychaudhuri, "Performance analysis of random-access packet switched code division multiple access channels," *IEEE Transactions on Communications*, vol. 29, pp. 895–901, June 1981.

[46] M. Joa-Ng and I. Lu, "Spread spectrum medium access protocol with collision avoidance in mobile ad-hoc wireless network," in *Proceedings of the Eighteenth Annual Joint Conference of the IEEE Computer and Communications Societies*, pp. 776–783, March 1999.

[47] G. Qiang, Z. Liu, S. Ishihara, and T. Mizuno, "CDMA-based carrier sense multiple access protocol for wireless LAN," in *Proceedings of the 2001 IEEE Vehicular Technology Conference*, vol. 2, pp. 1164–1168, 2001.

[48] D. P. Gerakoulis, T. N. Saadawi, and D. L. Schilling, "A channel access protocol for embedding CSMA on spread-spectrum packet radio networks," *IEEE International Conference on Communications*, pp. 199–204, 1988.

[49] F. L. Lo, T. S. Ng, and T. T. Yuk, "Performance analysis of a fully-connected, full-duplex CDMA ALOHA network with channel sensing and collision detection," *IEEE Journal on Selected Areas in Communications*, vol. 14, pp. 1708–1716, December 1996.

[50] M. V. Hedge and W. E. Stark, "Capacity of frequency-hop spread-spectrum multiple-access communication systems," *IEEE Transactions on Communications*, vol. 38, pp. 1050–1059, July 1990.

[51] U. Madhow and M. B. Pursley, "Limiting performance of frequency-hop random access," *IEEE Transactions on Information Theory*, vol. 36, pp. 322–333, March 1990.

[52] K. Cheun and K. Choi, "Performance of FHSS multiple access networks using MFSK modulation," *IEEE Transactions on Communications*, vol. 44, pp. 1514–1526, November 1996.

[53] A. S. Park, R. M. Buehrer, and B. D. Woerner, "Throughput performance of an FHMA system with variable rate coding," *IEEE Transactions on Communications*, vol. 46, pp. 521–532, April 1998.

[54] T. M. Cover and J. A. Thomas, *Elements of Information Theory*. New York: Wiley, 1991.

[55] S. Verdú, *Multiuser Detection*. New York: Cambridge University Press, 1998.

[56] K. S. Schneider, "Optimum detection of code division multiplexed signals," *IEEE Transactions on Aerospace and Electronic Systems*, vol. AES-15, pp. 181–185, January 1979.

[57] S. Verdú, "Minimum probability of error for asynchronous Gaussian multiple access channels," *IEEE Transactions on Information Theory*, vol. 32, pp. 85–96, January 1986.

[58] R. Kohno, M. Hatori, and H. Imai, "Cancellation techniques of co-channel interference in asynchronous spread spectrum multiple access systems," *Electronics and Communications in Japan*, vol. 66-A, no. 5, pp. 20–29, 1983.

[59] R. Lupas and S. Verdú, "Linear multiuser detectors for synchronous code-division multiple-access channels," *IEEE Transactions on Information Theory*, vol. 35, pp. 123–136, January 1989.

[60] R. Lupas and S. Verdú, "Near-far resistance of multiuser detectors in asynchronous channels," *IEEE Transactions on Communications*, vol. 38, pp. 496–508, April 1990.

[61] Z. Xie, C. K. Rushforth, and R. T. Short, "Multiuser signal detection using sequential decoding," *IEEE Transactions on Communications*, vol. 38, pp. 578–583, May 1990.

[62] M. K. Varanasi, "On multiuser detection in noncoherent CDMA channels," in *Proceedings of the 1991 IEEE Global Communications Conference*, pp. 846–850, December 1991.

[63] M. K. Varanasi and B. Aazhang, "Optimally near-far resistant multiuser detection in differentially coherent synchronous channels," *IEEE Transactions on Information Theory*, vol. 37, pp. 1006–1018, July 1991.

[64] M. K. Varanasi and S. Vasudevan, "Multiuser detectors for synchronous CDMA communications over non-selective Rician fading channels," *IEEE Transactions on Communications*, vol. 42, pp. 711–722, February/March/April 1994.

[65] S. Vasudevan and M. K. Varanasi, "Optimum diversity combiner based multiuser detection for time-dispersive Rician fading CDMA channels," *IEEE Journal on Selected Areas in Communications*, vol. 12, pp. 580–592, May 1994.

[66] Z. Zvonar and D. Brady, "Multiuser detection in single-path fading channels," *IEEE Transactions on Communications*, vol. 42, pp. 1729–1739, February/March/April 1994.

[67] Z. Zvonar and D. Brady, "Adaptive multiuser receivers with diversity reception for nonselective Rayleigh fading asynchronous CDMA channels," in *Proceedings of the 1994 IEEE Military Communications Conference*, vol. 3, pp. 982–986, 1994.

[68] Z. Zvonar and D. Brady, "Differentially coherent multiuser detection in asynchronous CDMA flat Rayleigh fading channels," *IEEE Transactions on Communications*, vol. 43, pp. 1252–1255, February/March/April 1995.

[69] Z. Zvonar and D. Brady, "Suboptimum multiuser detector for synchronous CDMA frequency-selective Rayleigh fading channels," in *Proceedings of the 1992 IEEE Global Communications Conference*, pp. 82–86, December 1992.

[70] Z. Zvonar and D. Brady, "On multiuser detection in asynchronous CDMA flat Rayleigh fading channels," in *Proceedings of the Third IEEE International Symposium on Personal, Indoor, and Mobile Radio Communications*, pp. 123–127, October 1992.

[71] H. C. Huang and S. C. Schwartz, "A comparative analysis of linear multiuser detectors for fading multipath channels," in *Proceedings of the 1994 IEEE Global Communications Conference*, pp. 11–15, November 1994.

[72] A. Klein and P. W. Baier, "Simultaneous cancellation of cross interference and ISI in CDMA mobile radio communications," in *Proceedings of the Third IEEE International Symposium on Personal, Indoor, and Mobile Radio Communications*, pp. 118–122, October 1992.

[73] A. Klein and P. W. Baier, "Linear unbiased data estimation in mobile radio systems applying CDMA," *IEEE Journal on Selected Areas in Communications*, vol. 11, pp. 1058–1066, September 1993.

[74] Z. Xie, R. T. Short, and C. K. Rushforth, "A family of suboptimum detectors for coherent multiuser communications," *IEEE Journal on Selected Areas in Communications*, vol. 8, pp. 683–690, May 1990.

[75] S. S. H. Wijayasuriya, G. H. Norton, and J. P. McGeehan, "A near-far resistant algorithm to combat the effects of fast fading in multi-user DS-CDMA systems," in *Proceedings of the Third IEEE International Symposium on Personal, Indoor, and Mobile Radio Communications*, pp. 645–649, October 1992.

[76] S. S. H. Wijayasuriya, G. H. Norton, and J. P. McGeehan, "A near-far resistant sliding window decorrelating algorithm for multi-user detectors in DS-CDMA systems," in *Proceedings of the 1992 IEEE Global Communications Conference*, pp. 1331–1338, December 1992.

[77] M. J. Juntti, "Linear multiuser detector update in synchronous dynamic CDMA systems," in *Proceedings of the Sixth IEEE International Symposium on Personal, Indoor, and Mobile Radio Communications*, vol. 3, p. 980, September 1995.

[78] U. Mitra and H. V. Poor, "Adaptive receiver algorithms for near-far resistant CDMA," *IEEE Transactions on Communications*, vol. 43, pp. 1713–1724, February/March/April 1995.

[79] Y. Bar-Ness and N. Sezgin, "Maximum signal-to-noise ratio data combining for one-shot asynchronous multiuser CDMA detector," in *Proceedings of the Sixth IEEE International Symposium on Personal, Indoor, and Mobile Radio Communications*, vol. 1, pp. 188–192, September 1995.

[80] Y. Miki, H. Andoh, and M. Sawahashi, "Coherent interference canceller system with pilot symbol-aided data estimation for forward-link and reverse-link channels of DS-CDMA," in *Proceedings of Wireless '95*, vol. 1, pp. 181–190, July 1995.

[81] M. Juntti, "Performance of multistage detector with one-shot decorrelating type first stage in an asynchronous DS/CDMA system," in *Proceedings of the Third IEEE International Symposium on Spread Spectrum Techniques and Applications*, vol. 1, pp. 157–161, July 1994.

[82]   F. Zheng and S. K. Barton, "One-shot near-far resistant CDMA detection in multipath fading channels—an O^3BPSK based system," in *Proceedings of the 1995 IEEE Vehicular Technology Conference*, vol. 1, pp. 489–493, July 1995.

[83]   Y. Bar-Ness and J. B. Punt, "An improved multi-user CDMA decorrelating detector," in *Proceedings of the Sixth IEEE International Symposium on Personal, Indoor, and Mobile Radio Communications*, vol. 3, p. 975, September 1995.

[84]   Y. Bar-Ness, Z. Siveski, and D. W. Chen, "Bootstrapped decorrelating algorithm for adaptive interference cancelation in synchronous CDMA communications systems," in *Proceedings of the Third IEEE International Symposium on Spread Spectrum Techniques and Applications*, vol. 1, pp. 162–166, July 1994.

[85]   D. S. Chen and S. Roy, "An adaptive multiuser receiver for CDMA systems," *IEEE Journal on Selected Areas in Communications*, vol. 12, pp. 808–816, June 1994.

[86]   K. Fukawa and H. Suzuki, "Orthogonalizing matched filter (OMF) detection for DS-CDMA mobile radio systems," in *Proceedings of the 1994 IEEE Global Communications Conference*, vol. 1, pp. 385–389, December 1994.

[87]   D. Jitsukawa and R. Kohno, "Multiuser equalization using multidimensional lattice filter for DS/CDMA," in *Proceedings of the 1995 Conference on Universal Personal Communications*, pp. 899–903, November 1995.

[88]   A. Kajiwara and M. Nakagawa, "Micro-cellular CDMA system with a linear multiuser interference canceller," *IEEE Journal on Selected Areas in Communications*, vol. 12, pp. 605–611, May 1994.

[89]   L. A. Rusch and H. V. Poor, "Multiuser detection techniques for narrowband interference suppression in spread spectrum communications," *IEEE Transactions on Communications*, vol. 43, pp. 1725–1737, February/March/April 1995.

[90]   P. Seite and J. Tardivel, "Adaptive equalizers for joint detection in an indoor CDMA channel," in *Proceedings of the 1995 IEEE Vehicular Technology Conference*, vol. 1, pp. 484–488, July 1995.

[91]   S. Verdú, "*Optimum multi-user signal detection*," Ph.D. dissertation, University of Illinois, Urbana-Champaign, 1984.

[92]   S. Verdú, "Adaptive multiuser detection," in *Proceedings of the Third IEEE International Symposium on Spread Spectrum Techniques and Applications*, vol. 1, pp. 43–50, July 1994.

[93]   A. Duel-Hallen, "Equalizers for multiple input/multiple output channels and PAM systems with cyclostationary input sequences," *IEEE Journal on Selected Areas in Communications*, vol. 10, pp. 630–639, April 1992.

[94]   A. Klein and P. W. Baier, "Equalizers for multi-user detection in code division multiple access mobile radio systems," in *Proceedings of the 1994 IEEE Vehicular Technology Conference*, vol. 2, pp. 762–766, June 1994.

[95] S. Moshavi, E. G. Kanterakis, and D. S. Schilling, "A new multiuser detection scheme for DS-CDMA systems," in *Proceedings of the 1995 IEEE Military Communications Conference*, vol. 2, pp. 518–522, November 1995.

[96] P. S. Kumar and J. Holtzman, "Power control for a spread spectrum system with multiuser receivers," in *Proceedings of the Sixth IEEE International Symposium on Personal, Indoor, and Mobile Radio Communications*, vol. 3, p. 955, September 1995.

[97] A. Duel-Hallen, "Decorrelating decision-feedback multiuser detector for synchronous code-division multiple-access channel," *IEEE Transactions on Communications*, vol. 41, pp. 285–290, February 1993.

[98] A. Duel-Hallen, "On suboptimal detection for asynchronous code-division multiple-access channels," in *Proceedings of the 26th Annual Conference on Information Sciences and Systems*, pp. 838–843, March 1992.

[99] A. J. Viterbi, "Very low rate convolutional codes for maximum theoretical performance of spread-spectrum multiple-access channels," *IEEE Journal on Selected Areas in Communications*, vol. 8, pp. 641–649, May 1990.

[100] P. Kempf, "On multi-user detection schemes for synchronous coherent CDMA systems," in *Proceedings of the 1995 IEEE Vehicular Technology Conference*, pp. 479–483, 1995.

[101] K. Puttegowda, G. Verma, S. Bali, and R. M. Buehrer, "On the effect of cancellation order in successive interference cancellation for CDMA systems," in *Proceedings of the 2003 IEEE Vehicular Technology Conference*, vol. 2, pp. 1035–1039, October 2003.

[102] R. M. Buehrer, "Equal BER performance in linear successive interference cancellation," *IEEE Transactions on Communications*, vol. 49, pp. 1250–1258, July 2001.

[103] P. Patel and J. Holtzman, "Analysis of a simple successive interference cancellation scheme in a DS/CDMA system," *IEEE Journal on Selected Areas in Communications*, vol. 12, pp. 796–807, June 1994.

[104] J. M. Holtzman, "Successive interference cancellation for direct sequence code division multiple access," in *Proceedings of the 1994 IEEE Military Communications Conference*, vol. 3, pp. 997–1001, October 1994.

[105] J. M. Holtzman, "DS/CDMA successive interference cancellation," in *Proceedings of the Third IEEE International Symposium on Spread Spectrum Techniques and Applications*, vol. 1, pp. 69–78, July 1994.

[106] R. M. Buehrer and R. Mahajan, "On the usefulness of power control with successive interference cancellation," *IEEE Transactions on Communications*, vol. 51, pp. 2091–2102, December 2003.

[107] G. Mazzini, "Equal BER with successive interference cancellation DS-CDMA systems on AWGN and Ricean channels," in *Proceedings of the Sixth IEEE International*